Mathematical Modeling in the Age of a Pandemic

Textbooks in Mathematics

Series editors:
Al Boggess, Kenneth H. Rosen

Real Analysis
With Proof Strategies
Daniel W. Cunningham

Train Your Brain
Challenging Yet Elementary Mathematics
Bogumil Kaminski, Pawel Pralat

Contemporary Abstract Algebra, Tenth Edition
Joseph A. Gallian

Geometry and Its Applications
Walter J. Meyer

Linear Algebra
What You Need to Know
Hugo J. Woerdeman

Introduction to Real Analysis, Third Edition
Manfred Stoll

Discovering Dynamical Systems Through Experiment and Inquiry
Thomas LoFaro, Jeff Ford

Functional Linear Algebra
Hannah Robbins

Introduction to Financial Mathematics
With Computer Applications
Donald R. Chambers, Qin Lu

Linear Algebra
An Inquiry-based Approach
Jeff Suzuki

Mathematical Modeling in the Age of a Pandemic
William P. Fox

https://www.routledge.com/Textbooks-in-Mathematics/book-series/CANDHTEXBOOMTH

Mathematical Modeling in the Age of a Pandemic

William P. Fox

CRC Press
Taylor & Francis Group
Boca Raton London New York

CRC Press is an imprint of the
Taylor & Francis Group, an **informa** business

A CHAPMAN & HALL BOOK

First edition published 2021
by CRC Press
6000 Broken Sound Parkway NW, Suite 300, Boca Raton, FL 33487-2742

and by CRC Press
2 Park Square, Milton Park, Abingdon, Oxon, OX14 4RN

© 2021 Taylor & Francis Group, LLC

CRC Press is an imprint of Taylor & Francis Group, LLC

Library of Congress Control Number: 2020951722

ISBN: 978-0-367-70312-7 (hbk)
ISBN: 978-0-367-68474-7 (pbk)
ISBN: 978-1-003-14563-9 (ebk)

Typeset in Garamond
by MPS Limited, Dehradun

Contents

Preface

Why this book and why now? You cannot watch or read any news these days without hearing about the models for COVID-19 or the testing that must occur to approve vaccines or treatments for the disease.

This book's purpose is to shed some light on the meaning and interpretations of many of the types of models that are or might be used in the presentation of analysis. From the virus itself and its infectious rates and death rates to explaining the process for testing a vaccine or eventually a cure, we build, present, and show model testing. We address, through examples, the concepts of social distancing as well as herd immunity. Understanding these concepts is essential to the entire modeling process of a pandemic.

The data for COVID-19 are always changing so we provide only a few snapshots using the most current data we had available while writing this book. We also attempt to add some validity to the models developed and used by showing how close to reality the models are to predicting "results" from previous pandemics such as the Spanish flu in 1918 and, more recently, the Hong Kong flu. We then apply those same models to Italy, New York City, and the United States as a whole. These locations were used because the data were easy to obtain.

Our hope is that this book provides valuable insights in understanding that modeling is a process. As more data become available, the better the model's accuracy. It is essential to understand that there are many assumptions that go into the development of each type of model. The assumptions influence the interpretation of the results.

It is also interesting to realize that regardless of the modeling approach, the results generally indicate approximately the same results. In the last chapter we present a brief summary and conclusion.

Let me briefly explain the flow of chapters. Chapter 1 presents an explanation of the modeling process as well as sets up some models we will revisit later. We start modeling with discrete dynamical systems. We choose this because the mathematics is in the modeling. The solution that we provide are numerical and graphical. We think this may eliminate the mathematical anxiety. Next, we go from a discrete dynamical system to a system of discrete dynamical systems. This is where we first encounter the SIR model, but we encounter it in terms of numerical and graphical results. In Chapters 3 and 4, we primarily revisit these same types of models, but this time in a continuous world. We also present in Chapter 3 some modeling on social distancing. In Chapter 6, we discuss some probabilistic models as well as nonlinear and logistic regression to build models for infections and deaths. Chapters 7 and 8 are inferential statistics models, primarily hypothesis testing. The hypothesis testing show and describe the methods and procedures for new medicines and vaccines that are currently being made and hopefully will be helpful. Chapter 9 quickly illustrates agent-based models. The agents are the infected person or persons within a population and the models show the results after some number of time period of results of the infection rates.

William P. Fox

About the Author

Dr. William P. Fox is currently a visiting professor of computational operations research at the College of William and Mary teaching both graduate and undergraduate courses. He is an emeritus professor in the Department of Defense Analysis at the Naval Postgraduate School, and he taught a three-course sequence in mathematical modeling for decision making. He received his BS from the United States Military Academy at West Point, New York, his MS in operations research from the Naval Postgraduate School, and his PhD in industrial engineering from Clemson University. He has taught at the United States Military Academy for twelve years until retiring from active military service, and at Francis Marion University where he was the chair of mathematics for eight years. He has many publications and scholarly activities including over 20 books, over 20 chapters of books and technical reports, over 150 journal articles, and over 150 conference presentations and mathematical modeling workshops. He has directed several international mathematical modeling contests through the Consortium of Mathematics and its Applications (COMAP): the HiMCM and the MCM. His interests include applied mathematics, optimization (linear and nonlinear), mathematical modeling, statistical models, model for decision making in business, industry, medical and government, and computer simulations. He was a member of INFORMS, the Military Application Society of INFORMS, the Mathematical Association of America, and the Society for Industrial and Applied Mathematics where in many of these societies he has held numerous positions.

Chapter 1

Modeling as a Process

1.1 Introduction

Why mathematical modeling? In the everyday news coverage of the present times, the words "model" and "models" are used as if everyone understands these concepts. The truth is most people do not. Not even every mathematician gets the idea of a mathematical model.

Consider the importance of modeling for decision making in business (B), industry (I), and government (G), BIG. BIG decision making is essential to success at all levels. We do not encourage "shooting from the hip" or simply flipping a coin to make a decision. At times it might appear that either of these are happening every day. We recommend good analysis that enables the decision maker to examine and question results in order to find the best alternative to choose or decision to make. This book presents, explains, and illustrates a modeling process and provides examples of decision making analysis throughout.

Let's describe a mathematical model as a mathematical description of a system using the language of mathematics. Why mathematical modeling? Mathematical modeling, business analytics, and operations research are all similar descriptions that represent the use of quantitative analysis to solve real problems. This process of developing such a mathematical model is termed mathematical modeling. Mathematical models are used in the natural sciences (such as physics, biology, earth science, and meteorology), engineering disciplines (e.g., computer science, systems engineering, operations research, and industrial engineering), and in the social sciences (such as business, economics, psychology, sociology, political science, and social networks).

The professional in these areas uses mathematical models all the time. A mathematical model may be used to help explain a system and to study the effects of different components, and to make *predictions* about behavior (Giordano, Fox, & Horton, 2013).

Mathematical models can take many forms, including, but not limited to, dynamical systems, statistical models, differential equations, optimization models, or game theoretic models. These and other types of models can overlap, where one output becomes the input for another similar or different model form. In many cases, the quality of a scientific field depends on how well the mathematical models developed on the theoretical side agree with results of repeatable experiments (Giordano et al., 2013). Any lack of agreement between theoretical mathematical models and experimental measurements leads to model refinements and better models. We do not plan to cover all the mathematical modeling processes here. We only provide an overview to the decision makers. Our goal is to offer *competent, confident problem solvers* for the 21st century. We suggest the books listed in the reference section in order to get familiar with many more modeling forms.

1.2 Background and the Modeling Process

1.2.1 Overview

Bender (1978) first introduced a process for modeling. He highlighted: formulate the model, outline the model, ask is it useful, and test the model. Others have expanded on this simple outlined process. Giordano, Fox, and Horton (2014) presented a six step process: identify the problem to be solved, make assumptions, solve the model, verify the model, implement the model, and maintain the model. Myer (1984) suggested some guidelines for modeling including formulation, mathematical manipulation, and evaluation. Meerschaert (1993) developed a five-step process: ask the question, select the modeling approach, formulate the model, solve the model, and answer the question. Albright (2010) subscribes mostly to concepts and processes described in previous editions of Giordano et al. (2014). Fox (2013) suggested an eight-step approach: understand the problem or question, make simplifying assumptions, define all variables, construct the model, solve and interpret the model, verify the model, consider the model's strengths and weaknesses, and implement the model.

Most of these pioneers in modeling have suggested similar starts in understanding the problem or question to be answered and making key assumptions to help enable the model to be built. We add in this process the need for sensitivity analysis and model testing to help insure we have a model that is performing correctly to answer the appropriate questions.

For example, student teams in the Mathematical Contest in Modeling, were building models to determine the all-time best college sports coach. One team picked a coach that coached less than a year, went undefeated for the remaining part of the year, and won their bowl game. Thus, his season was a perfect season. Their algorithm picked this person as the all-time best coach. Sensitivity analysis and model testing could have shown the fallacy to their model.

Someplace between the defining of the variables and the assumptions, we begin to consider the model's form and technique that might be used to solve the model. The list of techniques is boundless in mathematics and we will not list them here. Suffice, it to say it might be good to initially decide among the forms: deterministic or stochastic for the model, linear or nonlinear for the relationship of the variables, and continuous or discrete for the variables themselves.

For example, consider the following scenarios:

> Two persons are infected with COVID-19 and return home from abroad. How many people will these two ultimately infect if not quarantined for 14 days?

Consider locating emergency response teams within a county or region for response to COVID-19. Can we model location of ambulances to insure the maximum numbers of potential patients are covered by the emergency response teams? Can we find the minimum number of ambulances required?

You are a new emergency room (ER) administrator. You set new goals for your tenure as administrator. You analyze the current status of service to measure against your goals. Are you meeting patient demand? If not what can be done to improve service? You want to prevent catastrophic failure at your ER at your hospital.

You have many alternatives to choose from for your venture. You have certain decision criteria that you consider to use to help in making this future. Can we build a mathematical model to assist us in this decision?

You are worried about COVID-19. How do you get your hands, let alone your mind, such an invisible foe? We hear about models predicting numbers of infected, numbers of death, and the need for social distancing, and don't forget the face masks.

These are all events that we can model using mathematics. This chapter will help a decision maker understand what a mathematical modeler might do for them as a confident problem solver using the techniques of mathematical modeling. As a decision maker, understanding the possibilities and asking the key questions will enable better decisions to be made and lower the risks.

One topic we decide to address here is that of predictions. Often on the news we hear that the number of deaths from COVID-19 will be between two large numbers. Where did these numbers come from?

There are two possible answers here. One answer is the use of prediction intervals rather than just a point estimate (that most of you did in algebra) or for a model we ran it with two sets of distinct parameters, one for a best case and one set for a worst set. We will illustrate the first case here and the latter case in subsequent chapters,

> Before we start discussing the process, let's review a simple concept from algebra and its misunderstanding as a use in the modeling concept.

We are given the mathematical model, $y = f(x) = 2.1 x + 23.1$. We would like to predict the value when $x = 100$. Let's further assume that x is measured in days and the equation was developed using regression methods that were taught in college.

If we substitute $x = 100$ into our equation we obtain the following $y = f(100) = 2.1 (100) = 23.1 = 233.1$. This is completely correct and some mathematics textbooks don't even cover the required principle need and prediction intervals. This concept is used in the very back of most statistical books, even the ones for non-mathematics and science majors.

However, the concept is important to understand the range of numbers that are used. The concept is that 233.1 is not the answer, the answer lies in an interval. I'll present the equation as well as the interval.

A prediction interval for $f(x)$, the individual response of y for a specific x is given by

$$f(x) \pm t_{\alpha/2} \frac{s}{\sqrt{n}} \sqrt{1 + \frac{1}{n} + \frac{(x^* - \bar{x})^2}{\Sigma(x_i - \bar{x})^2}}$$

where x^* is the given value of x, n is the number of observation used to obtain the model, and $t_{\alpha/2}$ is the critical value with $n-2$ degrees of freedom.

Let's assume that we had 100 data points to generate our model, $t_{\alpha/2} = 1.98446$ if $\alpha = 0.05$. (We note $\alpha = 0.05$ is one that is often used.) Let's assume also that we know the mean of the x values, mean = 50.5, and we know $\Sigma(x_i - 50.5)^2$ as 83,325. We use $f(100) = 233.1$ in our process. Let's assume that the $SE = 8. = SE = \frac{standard\ deviation}{\sqrt{n}}$.

233.1 ± 1.98446(8) (1.004)
Upper bound is 249.039
Lower bound is 217.16
Thus our answer to this problem lies somewhere in the interval [217.16, 240.039].

This is a process that shows how we obtain intervals. We apply more of these prediction processes to more sophisticated models in Chapter 6.

Now, back to COVID-19.

In a recent article, the author mentions three reasons why we should not compare counties' coronavirus responses. We provide these below but stating the assumptions involved with these is what makes the results not comparable. It might even work by using a different metric for counting (AidadeOnlineDiscussion May 6, 2020).

1. Research has questioned the usefulness of death rates in judging countries' success in battling COVID-19.
2. There are high numbers of "excess deaths" in many countries, much higher than deaths officially attributed to the coronavirus.
3. Geography and demographics play a significant role in how countries have been affected.

Some countries have been praised for their swift responses to the coronavirus, while others have been criticized for doing too little, too late.

But an increasing focus on data is leading to questions over whether we are comparing like with like. After all, the way nations record tests and deaths varies, and there may be societal differences that need to be considered.

Read more: https://www.weforum.org/agenda/2020/05/compare-coronavirus-reponse-excess-deaths-rates/

These comments simply state that the assumptions of the models between countries and perhaps between states within the US might have different assumptions, which in return means that the models would be different.

1.2.2 The Modeling Process

We introduce the process of modeling and examine many different scenarios in which mathematical modeling can play a role.

The art of mathematical modeling is earned through experience of building and solving models. Modelers must be creative, innovative, inquisitive, and willing to try new techniques as well as being able to refine their models, if necessary. A major step in the process is "passing the common sense" test for use of the model.

In its basic form modeling consists of three steps:

Make assumptions
Do some "math"
Derive and interpret conclusions

To that end, one cannot question the mathematics and its solution but one can always question the assumptions used.

To gain insight we will consider one framework that will enable the modeler to address the largest number of problems. The key is that there is something *changing for which we want to know the effects and the results of the effects*. The problem might involve any system under analysis. The real-world system can be very simplistic or very complicated. This requires both types of real-world systems to be modeled with the same logical stepwise process.

Consider modeling an investment. Our first inclination is to use the equations about compound interest rates that we used in high school or college algebra. The compound interest formula calculates the value of a compound interest investment after 'n' interest periods.

$$A = P(1 - i)^n$$

where:

A = amount after n interest periods.
P = principal, the amount invested at the start.

i = the interest rate applying to each period.

n = the number of interest periods.

This is a continuous formula. Have you seen any banking institutions that give continuous interest? In our research, we have not. As a matter of fact at our local credit union they have a sign that says, "Money deposited after 10 AM does not get credited until the night after the deposit." This makes discrete compound interest on the balance a more compelling assumption.

A powerful paradigm that we use to model with discrete dynamical systems is:

future value = present value + change

The dynamical systems that we will study with this paradigm may differ in appearance and composition, but we will be able to solve a large class of these "seemingly" different dynamical systems with similar methods. In this chapter we will use iteration and graphical methods to answer questions about the discrete dynamical systems.

We use flow diagrams to help us see how the dependent variable changes (see figure 1.1b). These flow diagrams help to see the paradigm and put it into mathematical terms. Let's consider modeling the number of infections in Wuhan, China from January 23, 2020 until February 11, 2020. We have the data and we plot it. Clearly we see a curved trend increasing upward.

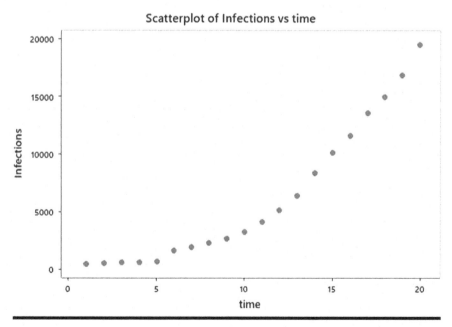

Figure 1.1a   Plot of infections versus time in Wuhan, China

Figure 1.1b   Flow diagram infections, Wuhan, China

We can estimate the growth rate using points to obtain a slope (Figure 1.1).

We use this change diagram to help build the discrete dynamical model. Let A (n) = the amount owed after n months. We note that susceptible population enters and after being infected, the population either recovers or dies.

We define the following variables:

$A (n + 1)$ = the number infected in the future
$A (n)$ = number currently infected

This is most likely exponential growth. We will model this later in more detail. Later in this chapter we briefly discuss or consider a drug dosage problem.

Figure 1.2 provides a closed loop process for modeling. Given a real-world situation like the one above, we collect data in order to formulate a mathematical model. This mathematical model can be one we derive or select from a collection of already existing mathematical models. Then we analyze the model that we used and reach mathematical conclusions about it. Next, we interpret the model and either makes predict about what has occurred or offer explanation as to why something has occurred. Finally, we test our conclusion about the real-world system with new data. We use sensitivity analysis of the parameters or inputs to see

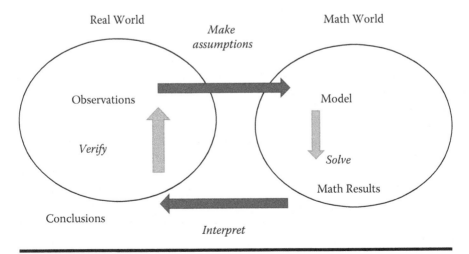

Figure 1.2   Modeling real-world systems with mathematics (see Albright, 2010)

how they affect the model. We may refine or improve the model to improve its ability to predict or explain the phenomena. We might even go back and re-formulate a new mathematical model.

1.2.3 The Pandemic as a Process

We will illustrate some mathematical models describing change in the real world. We will solve some of these models and analyze how good our resulting mathematical explanations and predictions are in context to the problem. The solution techniques that we employ take advantage of certain characteristics that the various models enjoy as realized through the formulation of the model.

When we observe change, we are often interested in understanding or explaining why or how a particular change occurs. Maybe, we need or want to analyze the effects under different conditions, or perhaps to predict what could happen in the future. Consider the firing of a weapon system or the shooting of a ball from a catapult, shown in Figure 1.3. Understanding how the system behaves in different environments under differing weather or operators, or predicting how well it hits the targets, are all of interest. For the catapult, the critical elements of the ball, the tension, and angle of the firing arm are found as important elements (see Fox, 2013). For our purposes, we will consider a mathematical model to be a mathematical construct designed to study a particular real-world system or behavior (see Giordano et al., 2013). The model allows us to use mathematical operations to reach mathematical conclusions about the model as illustrated in Figure 1.4. It is the arrow going from real-world system and observations to the mathematical model using the assumptions, variables, and formulation that is critical in the process.

We define a system as a set of objects joined by some regular interaction or interdependence in order for the complete system to work together. Think of a larger business with many companies working independently and interacting together to make the business prosper. Other examples might include a bass and

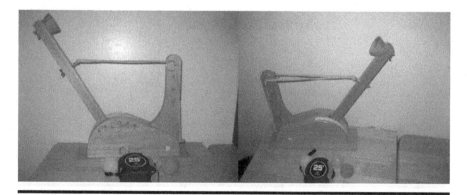

Figure 1.3 The catapult and balls

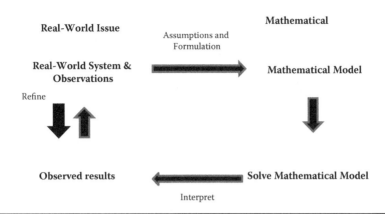

Figure 1.4 Modeling real-world systems with mathematics (adapted from Giordano et al., 2014)

trout population living in a lake, a communications system and cable TV, or a weather satellite orbiting the earth, delivering Amazon Prime packages, the US postal service mail or packages, locations of emergency services or computer terminals, or large companies' on-line customer buying systems. The person modeling is interested in understanding how a system works, what causes change in a system, and the sensitivity of the system to change. Understanding all these elements will help in building an adequate model to replicate reality. The person modeling is also interested in predicting what changes might occur and when these changes might occur.

Figure 1.5 suggests how we can obtain real-world conclusions from a mathematical model. First, observations identify the factors that seem to be involved in the behavior of interest. Often we cannot consider, or even identify, all the relevant factors, so we make simplifying assumptions excluding some of those factors (Giordano et al., 2014). Next, we determine what data are available and what variables they represent. We might build or test tentative relationships among the remaining identified factors. For example, if we were modeling the weight of a fish in a fish contest and we collected data on the length and girth of a fish, perhaps we would want to test the relationship of length and girth before we used both variables. This might give us a reasonable first cut at a model. We then solve the model and determine the reasonableness of the model's conclusions. Passing the "common sense" test is important. However, since these results apply only to the model, they may or may not apply to the actual real-world system in question. Simplifications were made in constructing the model and the observations upon which the model is based invariably contain errors and limitations. Thus, we must carefully account for these anomalies and test the conclusions of the model against real-world observations. If the model is reasonably valid, we can then draw inferences about the real-world behavior from the conclusions drawn from the

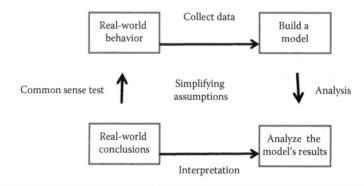

Figure 1.5 **In reaching conclusions about a real-world behavior, the modeling process is a closed system (adapted from Giordano et al., 2014)**

model. In summary, the critical elements are the modeling assumptions and variables. The mathematical model used is usually not questionable but why we used it might be questionable. Therefore, we have the following procedure for investigating real-world behavior and building a mathematical representation:

1. Observe the system and identify the factors and variables involved in the real-world behavior, possibly making simplifying assumptions as necessary.
2. Build initial relationships among the factors and variables.
3. Build the model and analyze the model's results.
4. Interpret the mathematical results both mathematically and in terms of the real-world system.
5. Test the model results and conclusions against real-world observations. Do the results and use of the model pass **the common sense** test? If not go back and remodel the system.

There are various kinds of models. A good mathematical modeler will build a library of models and to recognize various real-world situations to which they apply. Most models simplify reality. Generally, models can only approximate real-world behavior. Next, let's summarize a multi-step *process* for formulating a mathematical model.

1.2.4 Modeling Steps

An outline is presented as a procedure to help construct mathematical models. In the next section, we will illustrate this procedure with a few examples. We suggest a nine-step process.

These nine steps are summarized in Figure 1.6. These steps act as a guide for thinking about the problem and getting started in the modeling process. We

Step 1. Understand the decision to be made or the question asked.

Step 2. Make simplifying assumptions.

Step 3. Define all variables.

Step 4. Construct a model.

Step 5. Solve and interpret the model. Test the model. Do the results pass the *common sense test*?

Step 6. Verify the model. Validate, if possible.

Step 7. Identify the strengths and weaknesses as a reflection of your model.

Step 8. Sensitivity analysis or model testing.

Step 9. Implement and maintain the model for future use if it passes the common sense test.

Figure 1.6   Mathematical modeling process

choose these steps from the compilation of steps by other authors listed in additional readings and put them together in these nine steps.

We illustrate the process through an example. Consider building a model where we want to identify the spread of a contagious disease.

Step 1: Understand the decision to be made, the question to be asked, or the problem to be solved.

Understanding the decision is the same as identifying the problem to be solved. Identifying the problem to study is usually difficult. In real life no one walks up to you and hands you an equation to be solved. Usually, it is a comment like, "we need to make more money," or "we need to improve our efficiency." Perhaps, we need to make better decisions or we need all our units that are not 100% efficient to become more efficient. We need to be precise in our formulation of the mathematics to actually describe the situation that we need to solve. In our example, we want to identify the spread of a contagious disease to determine how fast it will spread within our region. Perhaps, we will want to use the model to answer the following questions:

1. How long will it take until one thousand people get the disease?
2. What actions may be taken to slow or eradicate the disease?

Step 2: Make simplifying assumptions.

Giordano et al. (2014, pp. 62–65), describe this well. Again we suggest starting by brain storming the situation. Make a list of as many factors, or variables, as you can. Now, we realize we usually cannot capture all these factors influencing a problem in our initial model. The task now is simplified by reducing the number of factors under consideration. We do this by making simplifying assumptions about the factors, such as holding certain factors as constants or ignoring some in the initial modeling phase. We might then examine to see if relationships exist between the remaining factors (or variables). Assuming simple relationships might reduce the complexity of the problem. Once you have a shorter list of variables, classify them as independent variables, dependent variables, or neither.

In our example, we assume we know the type disease, how it is spread, and the number of susceptible people within our region, and what type of medicine is needed to combat the disease. Perhaps we assume we know the size of population and the approximate number susceptible to getting the disease.

Falling Ladder Model
Let's consider the falling ladder model for an example of how assumptions change over time. Start with our related rates problem such as,

> "A 10-ft ladder is leaning against a house on flat ground. The house is to the left of the ladder. The base of the ladder starts to slide away from the house. When the base has slid to 8 feet from the house, it is moving horizontally at the rate of 2 ft/sec. How fast is the ladder's top sliding down the wall when the base is 8 feet from the house?"

Using calculus and starting with Pythagorean Theorem from a right triangle we end up with an equation

$$\frac{dy}{dt} = \frac{x}{y}\frac{dx}{dt}$$

From the problem, dx/dt = 2 ft/sec and x = 8 feet and $y = \sqrt{100 - 64}$ = 6 so 8/6 2 = 16/6 = 8/3 ft/sec.

But what if we want to know dy/dt when the ladder hits the ground. When the ladder hits the ground y = 0. Dividing by zero is not possible but indicates dy/dy = ∞. A force of ∞ would destroy the earth. So what happened? The math is correct but the mathematical assumptions behind this model have changed. As the ladder continuous to slide it separates from the wall as it closer to the ground. When that happens the end of ladder in the air falls according to the force of gravity.

So why this example? The assumptions behind the coronavirus are changing all the time. the infection rates, the death rates, the social distancing, the relaxation of social distancing all impact the modeling process.

Step 3: Define all variables.

It is critical to define all of your variables and provide the mathematical notation for each. Additionally if your variables have units include them as well. Include all variables, even those you might think that you will not initially use. Often we find we need these variables later in the refinement process.

In our example the variables of interest are the number of people currently infected, the number of people that are susceptible to the disease, and the number of people who recently recovered from the disease.

Step 4: Construct the model.

Using tools at your disposal or after learning new mathematical tools, you use your own creativity build a model that describes the situation and whose solution helps to answer important questions that have been asked. Generally three methods might be applied here. From first principles, your assumptions, and your variable list construct a useable mathematical model. From a data set, perform data analysis to examine for useful patterns that might suggest a useful model form such as a regression model. From research, take a model off the shelf and either use it directly or modify it appropriately for your use (a good discussion is found in Giordano et al. (2014).

In our example, we find we might be able to initially use the SIR model off the shelf.

Step 5: Solve and interpret the model.

We take the model that we have constructed in steps 1–4 and we solve it using mathematical tools. Often this model might be too complex or unwieldy so we cannot solve it or interpret it. If this happens, we return to steps 2–4 and simplify the model further. We can always try to enhance the model later. We also must insure that the model yields useable results for which the model was proposed. We will call this "passing the common sense test."

In our example the system of discrete dynamical system SIR equations has not closed form analytical solution. It does have graphical and numerical solutions that can be analyzed.

Step 6: Verify the model (see Giordano et al. (2014, pp. 63–64).

Before we use the model, we should test it out. There are several questions we must ask. Does the model directly answer the question or does the model allow for the answer to the questions to be answered? Is the model useable in a practical sense (can we obtain data to use the model)? Does the model pass the common sense test?

We provide an example to show this. We used a data set whose plot was reasonably a decreasing linear model. The correlation value was 0.94 which meant the data was strongly linear. We built the linear and used it to predict the value of a future y, where y had to be a positive value. The value of y, for our input x, was –23.5. So although many of the diagnostics were telling us that a linear model was adequate, the common sense test for using the model caused major refinements until we got a non-linear model that good diagnostics and passed the common sense tests (Fox, 2011).

Step 7: Strengths and weaknesses.

No model is complete without self-reflection of the modeling process. We need to consider not only what we did right but what we did that might be suspect as well as what we could do better. This reflection also helps in refining models in the future.

Step 8: Sensitivity analysis and model testing.

Sensitivity analysis is used to determine how "sensitive" a model is to changes in the value of the parameters of the model and to changes in the structure of the model. Parameter sensitivity is usually performed as a series of tests in which the modeler sets different parameter values to see how a change in the parameter causes a change in the dynamic behavior of the stocks. By showing how the model behavior responds to changes in parameter values, sensitivity analysis is a useful tool in model building as well as in model evaluation.

Sensitivity analysis helps to build confidence in the model by studying the uncertainties that are often associated with parameters in models. Many parameters in system dynamics models represent quantities that are very difficult, or even impossible to measure to a great deal of accuracy in the real world. Also, some parameter values change in the real world. Therefore, when building a system dynamics model, the modeler is usually at least somewhat uncertain about the parameter values he chooses and must use estimates. Sensitivity analysis allows him to determine what level of accuracy is necessary for a parameter to make the model sufficiently useful and valid. If the tests reveal that the model is insensitive, then it may be possible to use an estimate rather than a value with greater precision. Sensitivity analysis can also indicate which parameter values are reasonable to use in the model. If the model behaves as expected from real-world observations, it gives some indication that the parameter values reflect, at least in part, the "real world." Sensitivity tests help the modeler to understand dynamics of a system.

Experimenting with a wide range of values can offer insights into behavior of a system in extreme situations. Discovering that the system behavior greatly changes for a change in a parameter value can identify a leverage point in the model—a parameter whose specific value can significantly influence the behavior mode of the system.

In our SIR example, we would test how the changes in the parameters affect the solution.

Step 9: Implement and maintain the model **if it passes the common sense test**.

A model is pointless if we do not use it. The more user-friendly the model the more it will be used. Sometimes the ease of obtaining data for the model can dictate the model's success or failure. The model must also remain current. Often this entails updating data used for the model as well as updating the parameters used in the model.

1.2.5 Illustrative Examples

We now demonstrate the modeling process that was presented in the previous section. Emphasis is placed on problem identification and choosing appropriate (useable) variables in this section.

Example 1: Prescribed Drug Dosage

Scenario. Consider a patient that needs to take a newly marketed prescribed drug. To prescribe a safe and effective regimen for treating the disease, one must maintain a blood concentration above some effective level and below any unsafe level. How is this determined?

Understanding the Decision and Problem: Our goal is a mathematical model that relates dosage and time between dosages to the level of the drug in the bloodstream. What is the relationship between the amount of drug taken and the amount in the blood after time, t? By answering this question, we are empowered to examine other facets of the problem of taking a prescribed drug.

Assumptions: We should choose or know the disease in question and the type (name) of the drug that is to be taken. We will assume in this example that the drug is rythmol, a drug taken for the regulation of the heartbeat. We need to know or to find decaying rate of rythmol in the blood stream. This might be found from data that has been previously collected. We need to find the safe and unsafe levels of rythmol based upon the drug's "effects" within the body. This will serve as bounds for our model. Initially, we might assume that the patient size and weight has no effect on the drug's decay rate. We might assume that all patients are about the same size and weight. All are in good health and no one takes other drugs that affect the prescribed drug. We assume all internal organs are functionally properly. We might assume that we can model this using a discrete time period even though the absorption rate is a continuous function. These assumptions help simplify the model.

Example 2: Emergency Medical Response

The Emergency Service Coordinator (ESC) for a county is interested in locating the county's three ambulances to maximize the residents that can be reached within

Table 1.1   Average Travel Times from Zone *i* to Zone *j* in Perfect Conditions

	1	2	3	4	5	6
1	1	8	12	14	10	16
2	8	1	6	18	16	16
3	12	18	1.5	12	6	4
4	16	14	4	1	16	12
5	18	16	10	4	2	2
6	16	18	4	12	2	2

8 minutes in emergency situations. The county is divided into 6 zones and the average times required to travel from one region to the next under semi-perfect conditions are summarized in the following Table 1.1.

The population in zones 1, 2, 3, 4, 5 and 6 are given by the Table 1.2 below:

Understanding the Decision and Problem: We want better coverage and to improve the ability to take care of patients requiring to use an ambulance to go to a hospital. Determine the location for placement of the ambulances to maximize coverage within the predetermined allotted time.

Assumptions: We initially assume that time travel between zones is negligible. We further assume that the times in the data are averages under ideal circumstances.

Example 3: Medical Service Problem

The emergency room at a hospital is trying to improve customer satisfaction by offering better service. The ER wants the average patient wait to be less than

Table 1.2   Populations in Each Zone

1	50,000
2	80,000
3	30,000
4	55,000
5	35,000
6	20,000
Total	270,000

10 minutes and the average length of the queue (length of the line waiting) to be 5 minutes or fewer. The ER estimates about 150 patients per day. The existing arrival and service times are given in Tables 1.3 and 1.4.

Determine if the current ER service is satisfactory according to the guidelines. If not, determine through modeling the minimal changes for servers required to accomplish the ER's goal. We might begin by selecting a queuing model off the shelf to obtain some bench mark values.

Understand the Decision and Problem: The hospital wants to improve patient satisfaction. First we must determine if we are or are not meeting goals. Build a mathematical model to determine if the hospital is meeting its goals and if not come up with some recommendations to improve patient satisfaction.

Assumptions: Determine if the current patient service is satisfactory according to the ER's newest guidelines. If not, determine through modeling the minimal changes for servers required to accomplish the manager's goal. We might begin by selecting a queuing model off the shelf to obtain some bench mark values.

Table 1.3   Arrival Times

Time between Arrivals in Minutes	Probability
0	0.10
1	0.15
2	0.10
3	0.35
4	0.25
5	0.05

Table 1.4   Service Times

Service Time in Minutes	Probability
5	0.25
6	0.20
7	0.40
8	0.15

Table 1.5   Input and Outputs

Unit	Input #1	Input #2	Output #1	Output #2	Output #3
1	5	14	9	4	16
2	8	15	5	7	10
3	7	12	4	9	13

Example 4: Measuring Efficiency of Units

We have three major units where each unit has 2 inputs and 3 outputs as shown in Table 1.5.

Understand the Decision and Problem: We want to improve efficiency of our operation. We want to be able to find "best practices" to share. First, we have to measure efficiency. We need to build a mathematic model to examine efficiency of a unit based upon their inputs and outputs and be able to compare efficiency to other units.

Assumptions and Variable Definitions:
We define the following decision variables:

i_t= value of a single unit of output of *DMU i*, for i = *1, 2, 3*
w_i= cost or weights for one unit of inputs of *DMU i*, for i = *1, 2*
efficiency$_i$= (total value of *i*'s outputs)/(total cost of *i*'s inputs), for i = *1, 2, 3*

The following modeling initial assumptions are made:

1. No unit will have an efficiency more than 100%.
2. If any efficiency is less than 1, then it is inefficient.

Example 5: Wuhan, China

We have data per day of infections and deaths over a 20-day period. What can we say about those 20 days and what can we say about the future of infections and deaths (Table 1.6)?

Understand the Decision and Problem: We want to build and use a mathematical model of infections per day as well as build a model of deaths per day.

Assumptions Let's assume initially that Wuhan does not do anything to protect its people from the virus.

Table 1.6 Infections and Deaths, Wuhan, China (COVID-19 www.weforum.org)

Time	Infections	Deaths
1	425	17
2	495	23
3	572	38
4	618	45
5	698	63
6	1,590	85
7	1,905	104
8	2,261	129
9	2,639	159
10	3,215	192
11	4,109	240
12	5,142	270
13	6,384	300
14	8,351	371
15	10,117	432
16	11,618	535
17	13,603	602
18	14,982	665
19	16,902	738
20	19,558	810

Example 6: Risk Analysis for Homeland Security

Consider proving support to the Department of Homeland Security. The department only has so many assets and a finite amount of time to conduct investigations, thus priorities might be established. The risk assessment office has collected the data for the morning meeting shown in Table 1.7. Your operations research team must analyze the information and provide a priority list to the risk assessment team for that meeting.

Table 1.7 Risk Assessment Priority

Threat Alternatives\Criterion	Reliability of Threat Assessment	Approximate Associated Deaths (000)	Cost to Fix Damages in (Millions)	Location	Destructive Psychological Influence	Number of Intelligence-Related Tips
1-Dirty Bomb Threat	0.40	10	150	Urban dense	Extremely intense	3
2-Anthrax-Bio Terror Threat	0.45	.8	10	Urban dense	Intense	12
3-DC-Road & Bridge network threat	0.35	0.005	300	Urban & rural	Strong	8
4-NY subway threat	0.73	12	200	Urban dense	Very Strong	5
5-DC Metro Threat	0.69	11	200	Both Urban dense and rural	Very Strong	5
6-Major bank robbery	0.81	0.0002	10	Urban dense	Weak	16
7-FAA Threat	0.70	0.001	5	Rural dense	Moderate	15

Understand the Decision and Problem: There are more risks than we can possibly investigate. Perhaps if we rank these based upon useful criteria we can determine a priority for investigating these risks. We need to construct a useful mathematical model that ranks the incidents or risks in a priority order.

Assumptions: We have past decision that will give us insights into the decision maker's process. We have data only on reliability, approximate number of deaths, approximate costs to fix or rebuild, location, destructive influence, and number of intelligence gathering tips. These will be the criteria for our analysis ... The data is accurate and precise. We can convert word data into ordinal numbers.

Model: We could use multi-attribute decision making techniques for our model. We decide on a hybrid approach of AHP-TOPSIS. We will use AHP with Saaty's (1980) pairwise comparison to obtain the decision maker weights. We will also use the pairwise comparison to obtain numerical values for the criteria: location and destructive influence. Then we will use TOPSIS.

Example 7: Discrete SIR Models of Epidemics

Consider a disease that is spreading throughout the Unites States such as the new deadly flu. The CDC is interesting in know and experimenting with a model for this new disease prior to it actually becoming a "real" epidemic. Let us consider the population being divided into three categories: susceptible, infected, and removed. We make the following assumptions for our model:

- No one enters of leaves the community and there is no contact outside the community.
- Each person is either susceptible, S (able to catch this new flu); infected, I (currently has the flu and can spread the flu); or removed, R (already had the flu and will not get it again that includes death).
- Initially every person is either *S* or *I*.
- Once someone gets the flu this year they cannot get again.
- The average length of the disease is 2 weeks over which the person is deemed infected and can spread the disease.
- Our time period for the model will be per week.

The model we will consider is an off the shelf model, the SIR model (see Allman, 2004).

Let's assume the following definition for our variables.

$S(n)$ = number in the population susceptible after period n.
$I(n)$ = number infected after period n.
$R(n)$ = number removed after period n.

Let's start our modeling process with $R(n)$. Our assumption for the length of time someone has the flu is 2 weeks. Thus, half the infected people will be removed each week,

$$R(n + 1) = R(n) + 0.5\ I(n)$$

The value, 0.5, is called the removal rate per week. It represents the proportion of the infected persons who are removed from infection each week. If real data is available, then we could do "data analysis" in order to obtain the removal rate.

$I(n)$ will have terms that both increase and decrease its amount over time. It is decreased by the number that are removed each week, $0.5*I(n)$. It is increased by the numbers of susceptible that come into contact with an infected person and catch the disease, $a\ S(n)\ I(n)$. We define the rate, a, as the rate in which the disease is spread or the transmission coefficient. We realize this is a probabilistic coefficient. We will assume, initially, that this rate is a constant value that can be found from initial conditions.

Let's illustrate as follows. Assume we have a population of 1,000 students in the dorms. Our nurse found on 3 students reporting to the infirmary initially. The next week, 5 students came in to the infirmary with flu like symptoms. $I(0) = 3$, $S(0) = 997$. In week 1, the number of newly infected is 30.

$$5 = a\ I(n)S(n) = a(3) * (995)$$

$$a = 0.00167$$

Let's consider $S(n)$. This number is decreased only by the number that becomes infected. We may use the same rate, a, as before to obtain the model:

$$S(n + 1) = S(n) - a\ S(n)I(n)$$

Our coupled SIR model is

$$R(n + 1) = R(n) + 0.5I(n)$$
$$I(n + 1) = I(n) - 0.5I(n) + 0.00167I(n)S(n)$$
$$S(n + 1) = S(n) - 0.00167S(n)I(n)$$
$$I(0) = 3,\ S(0) = 997,\ R(0) = 0$$

The SIR model can be solved iteratively and viewed graphically. See will revisit this again in Chapter 3. In Chapter 5 we determine the worse of the flu epidemic occurs around week 8, at the maximum of the infected graph. The maximum number is slightly larger than 400, from the table in Chapter 7 it is approximated as 427. After 25 weeks, slightly more than 9 people never get the flu.

These examples will be solved in subsequent chapters.

Example 8: Application of Bayes' Theorem

Understand the Decision and Problem: More probabilistic information is better to improve our understanding. Assume we know at the end of a given day, the population numbers, the number infected, the number tested positive, and the number of deaths. Can we create a probabilistic model to assist in finding out probabilities?

Assumptions: We have past probabilities that might be used. But we need to know probabilities and conditional probabilities that are not readily available. We assume the Law of Total Probability and Bayes' Theorem might be used to gain insights.

Model: We assume the Law of Total Probability and Bayes' Theorem might be used to gain insights.

1.3 Technology

Most real-world problems that we have been involved in modeling solving require technology to assist the analyst, the modeler, and the decision maker. Microsoft Excel is available on most computers and represents a fairly good technological support for analysis for the average problems especially with *Analysis ToolPak* and the *Solver* installed. Other specialized software to assist analysts include: MATLAB, Maple, Mathematica, R, LINDO, LINGO, GAMS, as well as some additional add-ins for Excel such as the simulation package, Crystal Ball. Analysts should avail themselves to have access to as many of these packages as necessary. We only list those that we have used directly and realize that there are an endless quantity of models and technology available.

1.4 Conclusions

We have provided a clear and simple process to begin mathematical modeling in an applied situation requiring the stewardship of applied mathematics, operations research analysis, or risk assessment. We did not cover all the possible models but did highlight a few thorough illustrative examples. We emphasize that sensitivity analysis is extremely important in all models and should be accomplished prior to any decision being made. We show this in more detail in the chapters covering the techniques.

Chapter 2

Discrete Dynamical System Models

2.1 Introduction

Consider an inventory system where the management has restocking questions (from Meerschaert, 1999). A store has a limited number of 20-gallon aquarium tanks. At the end of each week the manager inventories and places orders. Store policy is to order 3 new 20-gallon tanks at the end of each week if all the inventory has been sold. If even 1 20-gallon aquarium tank remains in stock, no new units are ordered that week. This policy is based upon the observation that the store sells on average 1 of the 20-gallon aquarium tanks each week. Is this policy adequate to guard against lost sales?

Historical data are used to compute the probabilities of demand given the number of aquarium tanks on hand. From this analysis, we obtain the following change diagram, Figure 2.1.

Problem Statement: Determine the number of 20-gallon aquariums to order each week.

Assumptions and variables: Let n represent the number of time periods in weeks. We define

$A(n)$ = the probability that one aquarium is in demand in week n.
$B(n)$ = the probability that two aquariums are in demand in week n.
$C(n)$ = the probability that three aquariums are in demand in week n.

We assume that no other incentives are given to the demand for aquariums over this time frame. Can we build a model to analyze this inventory issue? We will solve this problem later in Chapter 3.

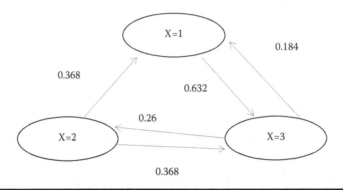

Figure 2.1   Flow diagram for inventory

2.2 Introduction to Modeling with Dynamical Systems and Difference Equations

Consider modeling infections due to COVID-19. Once you have looked at the data we might begin modeling. This process can be modeled as a dynamical system as susceptible becomes infected and infected can either recover or die.

We will develop a procedure how to proceed and solve problems in this chapter. We state up front that a nice way to examine all DDS problems is through iteration and graphs analysis.

2.3 Modeling Discrete Change

We are interested in modeling discrete *change*. Modeling with discrete dynamical systems employs a method to explain certain discrete behaviors or make long-term predictions. A powerful paradigm that we use to model with discrete dynamical systems is:

$$\textit{future value} = \textit{present value} + \textit{change}$$

The dynamical systems that we will study with this paradigm will differ in appearance and composition, but we will be able to solve a large class of these "seemingly" different dynamical systems with similar methods. In this chapter we will use iteration and graphical methods to answer questions about the discrete dynamical systems.

We will use flow diagrams to help us see how the dependent variable changes. These flow diagrams help to see the paradigm and put it into mathematical terms. Let's revisit a drug dosage problem. You are sick and you see a doctor. The doctor prescribes a medicine where you take 100 mg every four hours. Consider the flow diagram for the drug dosage, Figure 2.2, that depicts this situation.

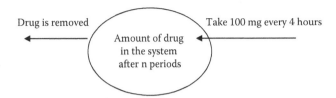

Figure 2.2 Flow diagram for a drug dosage problem

We use this flow diagram to help build the discrete dynamical model. Let A (n) = the amount of drug in your system after *n* periods. Notice that the arrow pointing into the circle is the drug being taken at the prescribed time periods. This increases the amount of the drug in the body. The arrow pointing out of the circle is amount of the drug that the body naturally processes out of the body that decreases the amount of drug in your systems.

a (n + 1) = the amount of drug after period n + 1

a (n) = amount currently in your body

a (n + 1) = a (n) − r a (n), where *r* is the rate in which your body naturally removes the drug

We will model dynamical systems that have only **constant coefficients**. A dynamical system with constant coefficients may be written in the form

$$a(n + 3) = b_2 a(n + 2) + b_1 a(n + 1) + b_0 a(n)$$

where b_0, b_1, and b_2 are arbitrary constants.

A **discrete dynamical system** is a "changing system," where the change of the system at each discrete iteration depends on (is related to) its previous state (or states) of the system. For a prescription drug dosage problem that we will also model, the amount of drug in the bloodstream after *n* hours depends on the amount of drug in the bloodstream after *n−1* hours. For financial matters, such as a mortgage balance or credit card balance, the amount you still owe after n months depends upon the amount you owed after *n−1* months. You will also find this process concerning states useful when we discuss Markov chains as an example of Discrete Dynamical Systems.

Example 2.1: Drug Dosage Problem

Suppose that a doctor prescribes that their patient takes a pill containing 100 mg of a certain drug every hour. Assume that the drug is immediately ingested into the bloodstream once taken. Also, assume that every hour the patient's body eliminates 25% of the drug that is in his/her bloodstream. Suppose that the patient had 0 mg of the drug in his/her bloodstream prior to taking the first pill. How much of the drug will be in his/her bloodstream after 72 hours?

Problem Statement: Determine the relationship between the amount of drug in the bloodstream and time.

Assumptions: The system can be modeled by a discrete dynamical system. The patient is of normal size and health. There are no other drugs being taken that will affect the prescribed drug. There are no internal or external factors that will affect the drug absorption rate. The patient always takes the prescribed dosage at the correct time.

Variables:

We define $a(n)$ to be the amount of drug in the bloodstream after period n, $n = 0, 1, 2...hours.$

Flow Diagram:

We create the input-output flow diagram as displaced in Figure 2.3.

Model Construction:

Let's define the following variables:

$$a(n + 1) = \text{amount of drug in the system in the future}$$

$$a(n) = \text{amount currently in system}$$

We define change as follows: change = dose − loss in system

$$change = 100 - .25 \ a(n)$$

So, Future = Present + Change is

$$a(n + 1) = a(n) - .25a(n) + 100$$

or

$$a(n + 1) = .75 \ a(n) + 100 \ \text{with} \ a(0) = 0.$$

Since the body loses 25% of the amount of drug in the bloodstream every hour, there would be 75% of the amount of drug in the bloodstream remaining every hour. After one hour, the body has 75% of the initial amount, 0 mg, to the 100 mg that is added every hour. So the body has 100 mg of drug in the bloodstream after one hour. After two hours the body has 75% of the amount of drug that was in the bloodstream after

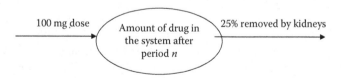

100 mg dose   Amount of drug in the system after period n 25% removed by kidneys

Figure 2.3   **Flow diagram for drugs in system**

one hour (100 mg), plus an additional 100 mg of drug added to the bloodstream. So there would be 175 mg of drug in the bloodstream after two hours. After three hours the body has 75% of the amount of drug that was in the bloodstream after two hours (175 mg), plus an additional 100 mg of drug added to the bloodstream. So there would be 231.25 mg of drug in the bloodstream after three hours.

An easy way to examine this is using MS-EXCEL. We will describe the process to iterate and graph the results shown in Figure 2.4.

Put the labels on cells, A3 and B3. In cell a4 put a 0 for time period 0 and 0 for the initial value of *a (0)*. In cell A5 type = *A4 + 1*, this increments the time period by 1. In cell B5, type = 0.75 ∗ b4 + 100, which represents the DDS. Then copy down for 30 time period.

Next, highlight the two columns, on the Command bar, go to INSERT, for the graphs choose the Scatterplot.

Interpretation of results: The DDS shows that the drug reaches a value where *change stops* and eventually the concentration in the bloodstream levels at 400 mg. If 400 mg is both a safe and effective dosage level then this dosage schedule is acceptable. We discuss this concept of change stopping (equilibrium) later in this chapter.

Example 2.2: Medicine, Such as a Tetanus Shot, over Time

You need a shot, such as tetanus, and receive it with 100% of the dose. We can measure the percent left in our body after n time periods. If we know we lost 0.10 of the drug every time period we can measure the decay over time.

Problem Statement: find a relationship between the amounts of the drug in our system as a function over time.

Assumptions: The rate in which the drug is depleted from our body is constant over the entire time period. No additional drug is added over the time period and the drug does not interact with any other drug or non-drug.

The following represents the decay of the drug over time:
Variables:
Let $d (n)$ = the amount of drug in our body after time period n where $n = 1, 2, 3,$
Flow Diagram:
The flow diagram is displayed in Figure 2.5.
Model Construction:
d $(n + 1)$ is the future
$d (n)$ = present

$$d (n + 1) = d (n) + 0.1 \ d (n)$$

or

$$d (n + 1) = 0.9 \ d (n)$$

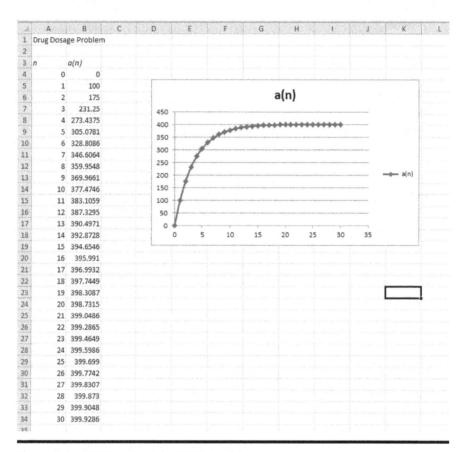

	A	B
1	Drug Dosage Problem	
2		
3	n	a(n)
4	0	0
5	1	100
6	2	175
7	3	231.25
8	4	273.4375
9	5	305.0781
10	6	328.8086
11	7	346.6064
12	8	359.9548
13	9	369.9661
14	10	377.4746
15	11	383.1059
16	12	387.3295
17	13	390.4971
18	14	392.8728
19	15	394.6546
20	16	395.991
21	17	396.9932
22	18	397.7449
23	19	398.3087
24	20	398.7315
25	21	399.0486
26	22	399.2865
27	23	399.4649
28	24	399.5986
29	25	399.699
30	26	399.7742
31	27	399.8307
32	28	399.873
33	29	399.9048
34	30	399.9286

Figure 2.4   EXCEL output for DDS model

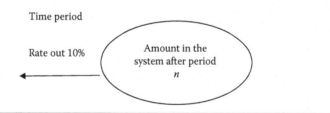

Time period

Rate out 10%   Amount in the system after period *n*

Figure 2.5   Flow diagram for money in a certificate

We know than initially, we had $1,000, so

$$d\,(0) = 100\% \; or \; d\,(0) = 1$$

Figure 2.6 shows that we will keep the drug decays and it appears to be approaching zero in the system.

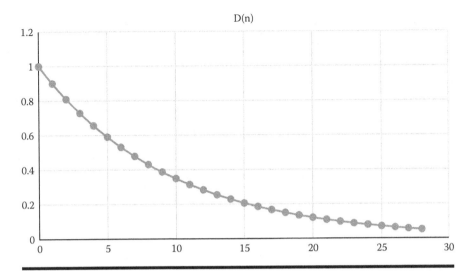

Figure 2.6   Plot of DDS for shot given 100% dose initially decays over time

Example 2.3: Drug Replenishment

Let's assume that in the previous example the drug, shot, is tetanus and we need a shot every 10 years. Why is this true? The shot decays at a certain rate and there is a range, upper bound and lower bound of drug in the system for which the drug remains effective.

Problem Identification: Build a model that relates the time with renewal of the tetanus shot.

Assumptions: The rate in which the drug is depleted from our body is constant over the entire time period. No additional drug is added over the time period and the drug does not interact with any other drug or non-drug.

The following represents the decay of the drug over time:

Variables:

Let $d(n)$ = the amount of drug in our body after time period n where $n = 1, 2, 3...$

Flow Diagram:

Variables:

Let $b(n)$ = amount in system after n time periods

Flow Diagram:

The flow diagram is displayed in Figure 2.7.

Model Construction:

$$b(n+1) = b(n) + .1\ b(n) + booster\ after\ x\ time\ periods$$

$$b(n+1) = 1.01\ b(n) - 880.87, \quad b(0) = 80,000$$

Figure 2.7 Flow diagram for shot example

Model Solution:

Graphical: First, we plot the DDS over the 30 time periods as displayed in Figure 2.8. We overlay the upper and lower effective bounds.

Example 2.4: Taking Prescribed Medication

Assume we are taking a medication every 8 hours. The pharmaceutical company has recommended every 8 hours for a specific dosage. The model they used is $D(n + 1) = D(n) - 0.5 * D(n)$ where n is measured in 8 hour increments. We present the iteration values and a plot to show that regardless whether or not the doctor gives you a large initial dose, a moderate initial dose, or small initial dose. We see that the model always goes toward 3, which is the level that is required to maintain a "cure" for the medical issue.

Cases		
I	*II*	*III*
6	3	1.5
4.5	3	2.25
3.75	3	2.625
3.375	3	2.8125
3.1875	3	2.9063
3.0938	3	2.9531
3.0469	3	2.9766
3.0234	3	2.9883
3.0117	3	2.9941
3.0059	3	2.9971
3.0029	3	2.9985

(*continued on next page*)

Cases		
I	*II*	*III*
3.0015	3	2.9993
3.0007	3	2.9996
3.0004	3	2.9998
3.0002	3	2.9999
3.0001	3	3

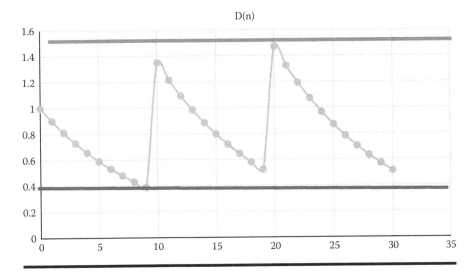

Figure 2.8 Shot cycle over time

A graphical result makes this clear. This is shown is Figure 2.9.

We will discuss the significance of the value of 3 next section.

However, you might see that is not exactly what happens. The body processes this drug at more than just 8-hour intervals.

We provide a plot based upon 1-hour intervals. We see the model setting around 3 as well, Figure 2.10.

2.4 Equilibrium Values and Long-Term Behavior

Equilibrium Values

Let's go back to our original paradigm,

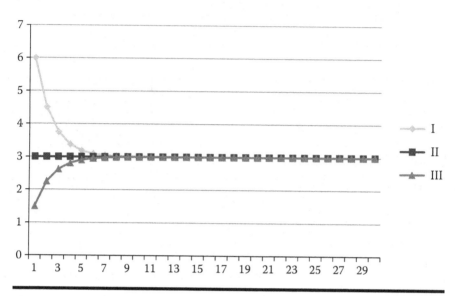

Figure 2.9   Model from different intial points converge to an equilibrium

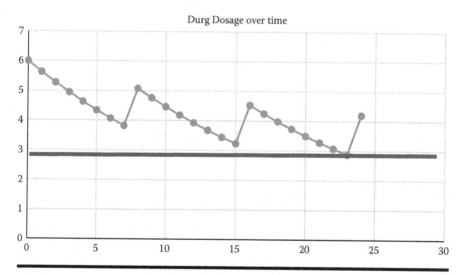

Figure 2.10   Drug dosage in 1-hour intervals over 24 hours

Future = Present + Change

When change stops, the change equals zero and future equals the present. The value for which this happens, if any, is the equilibrium value. This gives us a context for the concept of the equilibrium value.

We will define the Equilibrium Value (or fixed point) as the value where change stops. The value *a (n)* is an equilibrium value for the DDS, *a (n + 1)* = *f (a (n), a (n–1)...)* if for a value *a (0)* = A_0 all future values equal A_0.

Formally, we define the equilibrium value as follows:

> The number *ev* is called an **equilibrium value** or **fixed point** for a discrete dynamical system if *a (k)* = *ev* for all values of *k* when the initial value *a (0)* is set at *ev*. That is, *a (k)* = *ev* is a constant solution to the recurrence relation for the dynamical system.

Another way of characterizing such values is to note that a number a is an equilibrium value for a dynamical system *a(n + 1)* = *f(a(n), a(n–1),..., n)* if and only if a satisfies the equation *ev* = *f(ev, ev,..., e)*.

Using this definition, we can show that a linear homogeneous dynamical system of order 1 only has the value 0 as an equilibrium value.

In general, dynamical systems may have no equilibrium values, a single equilibrium value, or multiple equilibrium values. Linear systems have unique equilibrium values. The more non-linear a dynamical system is, the more equilibrium values it may have.

Not all DDS's have equilibrium values, and many DDS's that have equilibrium values that the system will never achieve. However, we already know that for a first order equation, if *a (0)* = *ev*, and every subsequent iteration value of the DDS is equal to *ev*, i.e., *a (k)* = *ev* for all value of *k*, then *ev* is an equilibrium value. For example, the DDS *a (n + 1)* = *2 a (n)* + 1 (Tower of Hanoi) has an equilibrium value of ev = –1. If we begin with *a (0)* = –1 and iterate, we get

$$a(1) = 2a(0) + 1 = 2(-1) + 1 = -2 + 1 = -1, \text{ so } a(1) = -1$$
$$a(2) = 2a(1) + 1 = 2(-1) + 1 = -2 + 1 = -1, \text{ so } a(2) = -1$$
$$a(3) = 2a(2) + 1 = 2(-1) + 1 = -2 + 1 = -1, \text{ so } a(3) = -1$$

etc.

So we say that when *a (0)* = –1, *a (k)* = –1 for all values of *k*, or in general, when *a (0)* = *ev*, then *a (k)* = *ev* for all values of *k*.

We can use this observation to find equilibrium values and to find out whether or not a DDS has an equilibrium value. Let's look at the DDS *a(n + 1)* = *2 a(n)* + 1, when *a(0)* = *ev*, *a(k)* = *ev* for all value of *k*, so that *a(1)* = *ev*, *a(2)* = *ev*, *a(3)* = *ev*,..., *a (n)* = *ev*, *a(n + 1)* = *ev*. Substituting *a (n)* = *ev* and *a (n + 1)* = *ev* into our DDS yields

$$a(n + 1) = 2 \ a(n) + 1$$

$$ev - 2 \ ev + 1$$

$$- ev = 1$$

$$ev = -1$$

So, the equilibrium value for the DDS is –1.

Now, let's consider the DDS $a(n + 1) = a(n) + 1$. Using our definition of equilibrium values, we write

$$a(n + 1) = a(n) + 1$$

$$ev = ev + 1$$

$$ev - ev = 1$$

$$0 = 1$$

The statement $0 = 1$ is not true, so the DDS **does not have an equilibrium value**.
Examples:
Consider the following DDS and find their equilibrium value, if they exist.

a. $a(n + 1) = 0.3\ a(n) - 10$
b. $a(n + 1) = 1.3\ a(n) + 20$
c. $a(n + 1) = 0.5\ a(n)$
d. $a(n + 1) = -.1\ a(n) + 11$

    Solutions:

a. $ev = .3\ ev - 10$
    $0.7\ ev = -10$
    $ev = -10/0.7 = -14.29$
b. $ev = 1.3\ ev + 20$
    $-0.3\ ev = 20$
    $ev = -20/0.3 = -66.66667$
c. $ev = 0.5ev,\ ev = 0$
d. $ev = -0.1\ ev + 11$

    $1.1\ ev = 11$
    $ev = 10$

We said that DDSs that have equilibrium values **may not ever attain** their equilibrium value, given some initial condition $a(0)$. For the "Tower of Hanoi," the equilibrium value was –1. Suppose that we begin iterating the DDS with an initial value $a(0) = 0$:

$$a(1) = 2(0) + 1 = 3$$
$$a(2) = 2(3) + 1 = 7$$
$$A(3) = 2(7) + 1 = 15$$
$$A(4) = 2(15) + 1 = 31$$
$$(0) = 0: \text{etc.}$$

The values continue to get larger and will never reach the value of −1. Since *a (n)* represents the number of moves of the disks, the value of −1 makes no real sense in the contest of the number of disk to move.

We will study equilibrium values in many of the applications of discrete dynamical systems. In general, a linear, non-homogeneous discrete dynamical system, where the non-homogeneous part is a constant, will have an equilibrium value. (Can you find any exceptions?) Linear, homogeneous discrete dynamical systems will have an equilibrium value of zero. (Why?)

A Graphical Approach to Equilibrium Values

We can examine the plot of the iterations using technology. If the values reach a specific value and remain constant then that value is an equilibrium value (change has stopped).

As seen, dynamical systems often represent real-world behavior that we are trying to understand. At times, we want to predict future behavior and gain deeper insights into how to influence or alter the behavior. Thus, we have great interest in the predictions of the model. How does it change in the future?

Models of the form: a (n + 1) = r a (n), r is a constant

Let's look at our savings account where we invest $1,000 at 12% a year compounded monthly.

$$a(n + 1) = 1.01 \, a(n), \quad a(0) = 1000$$

This sequence, with r > 1, grows without bound. The graph in Figure 2.11 suggests that there is no equilibrium value (where the graph levels out to a constant value). Analytically, *ev = 1.1ev + 1,000, ev = −909.09*. There is an equilibrium value of −909.09 but that value will never be reached in our savings account problem.

Now, what happens if r is less than 0? We replace r = 1.01 with r = −1.01 in the previous example. First, we can analytically solve for the equilibrium value.

$$ev = -1.01 \, ev + 1000$$

$$2.01 \, ev = 1000$$

$$ev = 497.5124378$$

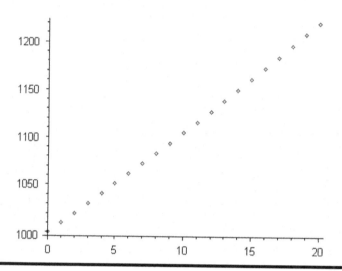

Figure 2.11   Saving account revisit

The definition implies that if we start at *a (0)* = *497.5124378* we stay there forever. If we plot the solution we note the oscillations between positive and negative numbers, each growing without bound as the oscillations fan out as shown in Figure 2.12. Although there is an equilibrium value, the solution to our example does not tend toward this equilibrium value.

Let's examine values of r, 0 < r < 1. Let's take a look at a drug dosage model, a (n + 1) = 0.5 a (n) with a (0) = 20, where half of what is in the system is discarded each time period.

The equilibrium value is zero.

Models of the form a (n + 1) = r a (n) + b, where r and b are constants

Let's return to our drug dosage problem and consider adding the constant dosage each time period (time periods might be 4 hours). Our model is a (n + 1) =0.5 a (n) +16 mg. We will also assume that there is an initial dosage applied prior to beginning the regime. We will let these initial values be as follows as plot the results in Figure 2.13 a-c.

Regardless of the starting value, the future terms of *a (n)* approach 32. Thus, 32 is the equilibrium value. We could have solved for this algebraically as well.

$$a (n + 1) = 0.5 \ a (n) + 16$$

$$ev = 0.5 \ ev + 16$$

$$0.5 \ ev = 16$$

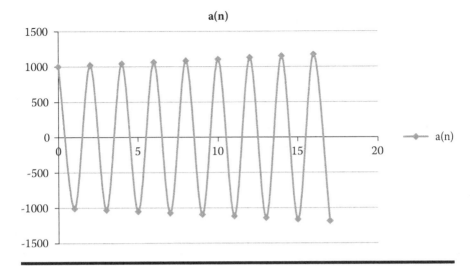

Figure 2.12   DDS graph using r = 1.01 and r −1.01

$$ev = 32$$

Another method of finding the equilibrium values involves solving the equation $a = r\,a + b$ and solving for a *(where a is ev)* we find:

$$a = \frac{b}{1 - r}, \quad \text{if } r \neq 1.$$

Using this formula in our previous example, the equilibrium value is

$$a = 16/(1 - .5) = 32.$$

2.4.1 Stability and Long-Term Behavior

For a dynamical system, $a\,(n + 1)$ with a specific initial condition, $a\,(0) = a_0$, we have shown that we can compute $a\,(1)$, $a\,(2)$, and so forth. Often these particular values are not as important as the long term behavior. By long term behavior, we refer to what will eventually happen to $a\,(n)$ for larger values of n. There are many types of long term behavior that can occur with DDS, we will only discuss a few here.

If the $a(n)$ values for a DDS eventually get close to the equilibrium value, ev, no matter the initial condition, then the equilibrium value is called a **stable equilibrium value** or **an attracting fixed point**.

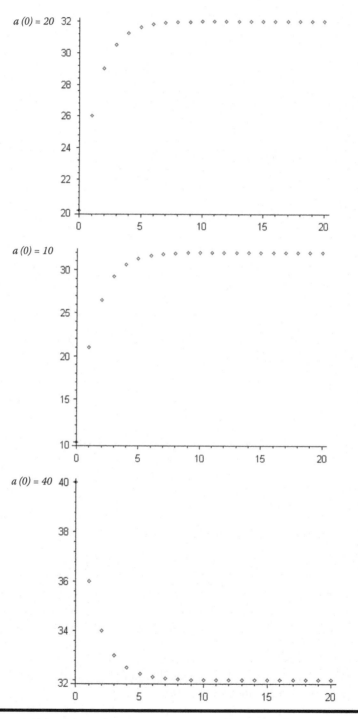

Figure 2.13 **Plot of drugs in our systems with different starting conditions**

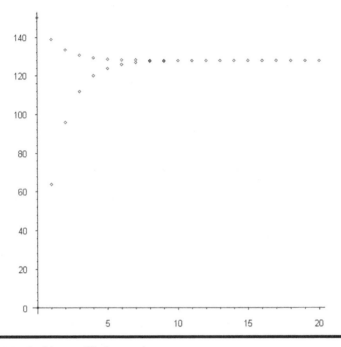

Figure 2.14 Stable equilibrium value

Example: Consider the following DDS

$A\ (n + 1) = 0.5\ A\ (n) + 64$, with initial conditions A (0) = 0 or A (0) = 150, the *eV* is 128. The ev is stable as shown in Figure 2.14.

Notice that both sequences are converging on 128 as the attracting fixed point or equilibrium value.

Example: Consider the DDS, $A\ (n + 1) = -.5\ A\ (n) + 10$ with *ev* = 6.66667 With A (0) =100, this is the plot of behavior, Figure 2.15.

Example: Consider the DDS for the financial model, $A\ (n + 1) = 1.1\ A\ (n) + 100$, A (0) = 100.

The *ev* value is –1,000. If the DDS ever achieves an input of –1,000 then the systems stays at –1,000 forever. But when we start at typical values like $100 to begin the process we find the values tend to move away from the *ev*. When this occurs, we say the ev is unstable or a repelling fixed point, Figure 2.16.

The values tend to increase over time and never move toward –1,000. Therefore, the *ev* is unstable.

Often we characterize the long-term behavior of the system in terms of its stability. If a DDS has an equilibrium value and if the DDS tends to the equilibrium value from starting values near the equilibrium value, then the DDS is said to be stable.

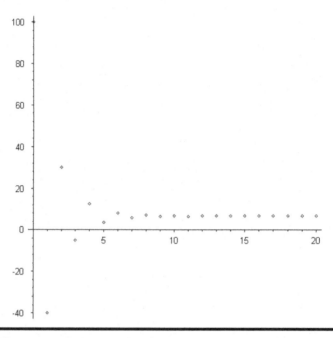

Figure 2.15 Plot of *A (n + 1) = −.5 A (n) +10* with different starting values showing stability of the equilibrium value

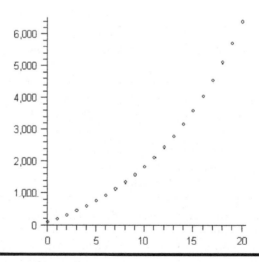

Figure 2.16 Unstable equilibrium value as the solution tends to move away from the equilibrium value

Thus, for the dynamical system $a (n + 1) = r a (n) + b$, where $b \neq 0$ we provide Table 2.1 to show the stability results.

If $r \neq 1$, an equilibrium exists at $a = b/ (1-r)$.
If $r = 1$, no equilibrium value exists.

Relationship to Analytical Solutions
If a discrete dynamical system has an ev value, we can use the *ev* value to find the analytical solution.
Recall the mortgage example from Section 2.2,

$$B (n + 1) = 1.00541667 \ B (n) - 639.34,$$

$$B (0) = 73,395$$

The *ev* value is found as *1,18,031.9274.*
The analytical solution may be found using the following form:

$$B (k) = (1.00541667^k) C + D \ \text{where} \ D \ \text{is the} \ ev.$$

$$B (k) = (1.00541667^k) C + 118031.9274, \ \ B (0) = 73.395$$

$$\text{Since} \ B (0) = 73395 = 1.00541667^0 (C) + 118031.9274$$

$$C = -44636.92736$$

Table 2.1   Stability of $a (n + 1) = r a (n) + b$, Where $b \neq 0$

Value of r	DDS Form	Equilibrium	Stability of Solution	Long-term Behavior
$r = 0$	$a(n + 1) = b$	b	Stable	Stable equilibrium
$r = 1$	$a(n + 1) = a(n) + b$	None	Unstable	
$r < 0$	$a(n + 1) = r*a(n) + b$	$b/(1-r)$	Depends on $|r|$	Oscillations
$|r| <1$	$a(n + 1) = r*a(n) + b$	$b/(1-r)$	Stable	Approaches $b/(1-r)$
$|r| >1$	$a(n + 1) = r*a(n) + b$	$b/(1-r)$	Unstable	Unbounded

Thus, the closed form model is,

$$B(k) = -44636.92737(1.00541667^k) + 118031.9274$$

Let's assume we did not know the payment was 639.34 per month. We could use the analytical solutions to help find the payment.

$$B(k) = (1.00541667^k)C + D$$

We build a system of two equations and two unknowns.

$$B(K) = (1.00541667^k)C + D$$
$$B(0) = 73395 = C + D$$
$$B(180) = 0 = 1.00541667^{180}C + D$$
$$C = -44638.70, D = 118033.7$$
$$B(K) = -44638.70(1.00541667)^k + 118033.7$$

D represents the equilibrium value and we accepted some round-off error. From our model form:

$$B(n + 1) = 1.00541667 \ B \ (n) - P, \quad \text{we can find } P.$$

Solving analytically for the equilibrium value,

$$X - 1.00541667X = -P$$

$$X = P/.00541667$$

$$X \text{ is } 118033.70 \text{ so}$$

$$118033.70 = P/.00541667$$

$$P = 639.34$$

Example 2.4 was a prescribed drug dosage problem. The *ev* is 3, which we saw through iteration and graphical means. We can compute it as

$$ev = ev - .5 * ev + 1.5$$

$$.5 \ ev = 1.5$$

$$ev = 3$$

Furthermore, our graph shows us that we start close to the equilibrium value we always return to the equilibrium value. That makes the equilibrium value stable.

2.5 Modeling Nonlinear Discrete Dynamical Systems

In this section we build nonlinear discrete dynamical systems to describe the change in behavior of the quantities we study. We also will study systems of DDS to describe the changes in various systems that act together in some way or ways. We define a nonlinear DDS-- If the function of $a\ (n)$ involves powers of $a\ (n)$ (like $a^2(n)$), or a functional relationship (like $a\ (n)/a\ (n * 1)$), we will say that the discrete dynamical system is **nonlinear**. A **sequence** is a function whose domain is the set of non-negative integers (n = 0, 1, 2...). We will restrict our model solution to the numerical and graphical solutions. Analytical solutions may be studied in more advanced mathematics courses.

Example 2.5: Population Growth: Growth of a Yeast Culture (Giordano et al., 2013)

We often model population growth by assuming that the change in population is directly proportional to the current size of the given population. This produces a simple, first order DDS similar to those seen earlier. It might appear reasonable at first examination, but the long-term behavior of growth without bound is disturbing. Why would growth without bound of a yeast culture in a jar (or controlled space) be alarming?

There are certain factors that affect population growth. These include resources (food, oxygen, space, etc.) These resources can support some maximum population. As this number is approached, the change (or growth rate) should decrease and the population should never exceed its resource supported amount.

Problem Statement: Predict the growth of yeast in a controlled environment as a function of the resources available and the current population.

Assumptions and Variables:

We assume that the population size is best described by the weight of the biomass of the culture. We define $y\ (n)$ as the population size of the yeast culture after period n. There exists a maximum carrying capacity, M, that is sustainable by the resources available. The yeast culture is growing under the conditions established.

Model:

$Y\ (n + 1) = y\ (n) + k\ y\ (n)\ (M\text{-}y\ (n))$ where

$Y\ (n)$ is the population size after period n

n is the time period measured in hours

k is the constant of proportionality

M is the carrying capacity of our system

Table 2.2 shows the data collected on the culture.

Table 2.2   Growth of Yeast in a Culture

	A	B	C	D
1	Time	Yeast Biomass, P(n)	P(n+1)-P(n)	P(n)*(1000-P(n))
2	0	14.496	13.137	14285.86598
3	1	27.633	16.157	26869.41731
4	2	43.79	27.482	41872.4359
5	3	71.272	36.089	66192.30202
6	4	107.361	72.48	95834.61568
7	5	179.841	83.805	147498.2147
8	6	263.646	124.877	194136.7867
9	7	388.523	141.034	237572.8785
10	8	529.557	136.353	249126.3838
11	9	665.91	109.173	222473.8719
12	10	775.083	70.064	174329.3431
13	11	845.147	53.001	130873.5484
14	12	898.148	52.246	91478.1701
15	13	950.394	17.214	47145.24476
16	14	967.608	15.553	31342.75834
17	15	983.161	7.248	16555.44808
18	16	990.409	5.587	9499.012719
19	17	995.996	3.314	3987.967984
20	18	999.31		
21				

We plot the time versus the Biomass, P (n), and see that as time increases it appears to stabilize at 1,000, Figure 2.17.

Next, we plot $y(n+1)-y(n)$ versus $y(n)(1,000-y(n))$ to find the slope, k, is approximately *0.000544. (shown in* Figure 2.18) With $k = 0.000544$ and the carrying capacity in biomass is *1,000*. This model is

$$y(n+1) = y(n) + .00082\ y(n)(1000 - y(n))$$

Again, this is nonlinear because of the $y^2(n)$ term. The solution iterated (there is no closed form analytical solution for this equation) from an initial condition, biomass, of 14.496:

The model and plot, Figure 2.19, shows stability in that the population (biomass) of the yeast culture approaches 1,000 as n gets large. Thus, the population is eventually stable at approximately 1,000 units.

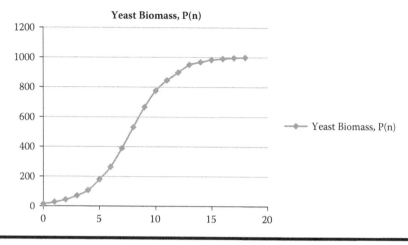

Figure 2.17   Biomass as a function of time

This same type model could be used per person to model the growth of virus cells provided we had data.

Example 2.6: Spread of a Contagious Disease (Fox, 2013)

Suppose that there are 1,000 students in a college dormitory, and some students have been diagnosed with COVID-19, a highly contagious disease. The health center wants to build a model to determine how fast the disease will spread.

Problem Identification: Predict the number of students affected with meningitis as a function of time.

Assumptions and Variables: Let m (n) be the number of students affected with meningitis after n days. We assume all students are susceptible to the disease. The possible interactions of infected and susceptible students are proportional to their product (as an interaction term).

The model is,

$$m(n+1) - m(n) = k \ m(n)(1000 - m(n)) \text{ or}$$

$$m(n+1) = m(n) + k \ m(n)(1000 - m(n))$$

Two students returned from spring break with COVID-19. The rate of spreading per day is characterized by $k = 0.0090$. It is assumed that a vaccine can be in place and students vaccinated within 1–2 weeks.

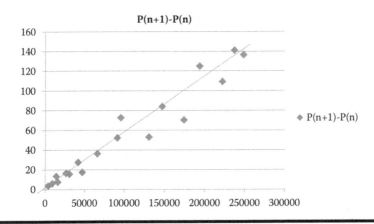

Figure 2.18 Plot of *y (n + 1)-y (n)* versus *y (n) (1,000-y (n))*

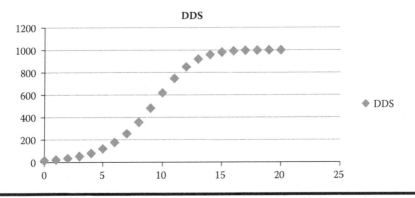

Figure 2.19 Plot of DDS from growth of a yeast culture

$$m(n + 1) = m(n) + 0.00090\ m(n)(1000 - m(n))$$

This is decided in Figure 2.20.

Interpretation: The results show that most students will be affected within 2 weeks. Since only about 10% will be affected within one week, every effort must be made to get these students quarantined and isolated (for 14 days) or until a vaccination is available at the school and get all the students vaccinated within one week.

Example 2.7: COVID-19 in Wuhan, China (data from Jan and Feb 2020)

We apply the growth of a yeast culture DDS model to the number of infected in Wuhan, China.

Time	Infections	P(n +1) – P(n)	p (n)*(11m – P(n))	P(n +1) – P(n)	Scaled Values
1	425	70	4,674,819,375	70	467.4819
2	495	77	5,444,754,975	77	544.4755
3	572	46	6,291,672,816	46	629.1673
4	618	80	6,797,618,076	80	679.7618
5	698	892	7,677,512,796	892	767.7513
6	1,590	315	1.7487E+10	315	1,748.747
7	1,905	356	2.0951E+10	356	2,095.137
8	2,261	378	2.4866E+10	378	2,486.589
9	2,639	576	2.9022E+10	576	2,902.204
10	3,215	894	3.5355E+10	894	3,535.466
11	4,109	1,033	4.5182E+10	1,033	4,518.212
12	5,142	1,242	5.6536E+10	1,242	5,653.556
13	6,384	1,967	7.0183E+10	1,967	7,018.324
14	8,351	1,766	9.1791E+10	1,766	9,179.126
15	10,117	1,501	1.1118E+11	1501	11,118.46
16	11,618	1,985	1.2766E+11	1,985	12,766.3
17	13,603	1,379	1.4945E+11	1,379	14,944.8
18	14,982	1,920	1.6458E+11	1,920	16,457.75
19	16,902	2,656	1.8564E+11	2,656	18,563.63
20	19,558				

We plot the values to obtain a slope, the proportionality constant.

We used linear regression to find the slope as 0.143312. This is shown in Figure 2.21.

Our model is p *(n + 1) = 0.143312 p (n) (11m - p (n))*.

We use this model to estimate infections in an additional 15 days. The plot is shown as Figure 2.22.

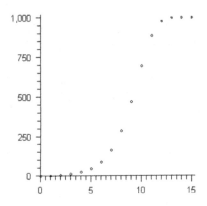

Figure 2.20   Plot of DDS for the spread of a disease

Example 2.8: Spread of a Pandemic in the United States

Suppose that there are 334,000,000 people in the United States, and some people have been diagnosed with COVID-19, a highly contagious disease. The CDC wants to build a model to determine how fast the disease will spread.

Problem Identification: Predict the number of people affected with COVID-19 as a function of time.

Assumptions and Variables: Let *m (n)* be the number of students affected with COVID-19 after *n* days. We assume all people are susceptible to the disease. The possible interactions of infected and susceptible students are proportional to their product (as an interaction term). We assume that there is quarantine or isolation in the general public as well as no testing.

The graph of the actual data, Figure 2.23, (COVID-19 Modeling and Predictions, M. Bhatnager) suggests the growth rate is approximately 100 per day.

The model is,

$$m(n + 1) - m(n) = k \ m(n)(334{,}000{,}000 - m(n)) \ \text{or}$$

$$m(n + 1) = m(n) + 1/10000 \ m(n)(334{,}000{,}000 - m(n))$$

Two travelers returned from a holiday with COVID-19. The rate of spreading per day is characterized by *k = 1/10,000.*

$$m(n + 1) = m(n) + 1/10000 \ m(n)(334{,}000{,}000 - m(n))$$

This simple model shows that after 60 days, about 1.9 million people are infected. This is seen in Figure 2.24.

If there are no quarantines or isolations then after about 1 year, we will have over 11 million infected. Now, we are only talking about infections. We are not talking about deaths from the COVID-19 disease.

Plot of proportionality argument

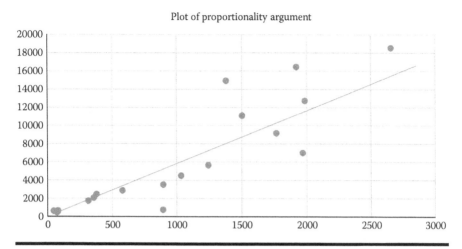

Figure 2.21   Proportionality plot for Wuhan, China

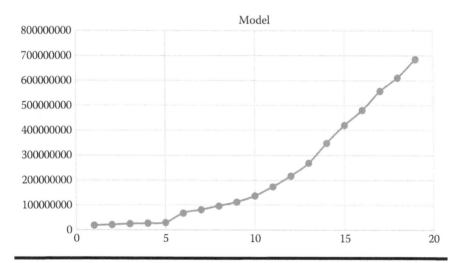

Figure 2.22   Plot of Wuhan, China model

Example 2.9: Spread in a Specific Location, New York City

Suppose that there are 8.77 million people in New York City and some people have been diagnosed with COVID-19, a highly contagious disease. The CDC wants to build a model to determine how fast the disease will spread.

Problem Identification: Predict the number of people affected with COVID-19 as a function of time.

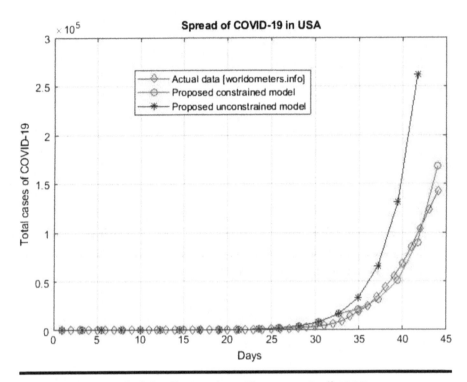

Figure 2.23 Spread of the disease (from Bhatnager, April 2020)

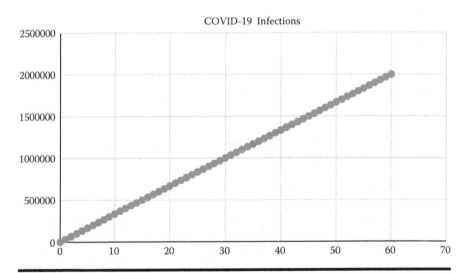

Figure 2.24 US COVID-19 infections

Assumptions and Variables: Let m (n) be the number of students affected with COVID-19 after n days. We assume all people are susceptible to the disease. The possible interactions of infected and susceptible people are proportional to their product (as an interaction term). We assume that there is quarantine or isolation in the general public as well as no testing. We assume we will average the cases per day from Figure 2.25 and assume this value is 3,000.

The model is,

$$m(n + 1) - m(n) = k \ m(n)(8{,}770{,}000 - m \ (n)) \ \text{or}$$

$$m(n + 1) = m(n) + 1/ \ 3000 \ m(n)(8{,}770{,}000 - m(n))$$

Two travelers returned from a holiday with COVID-19. The rate of spreading per day is characterized by $k = 1/3{,}000$.

$$m(n + 1) = m(n) + 1/3000 \ m(n)(8{,}770{,}000 - m(n))$$

Using this simple model, results seen in Figure 2.26, we predict 173,689 cases.

According to NY Health there have been 174,709 actual diagnosed cases. Our model has an error rate of –0.583%.

One thing that we have seen to date is that something needs to be done to slow the progression of the virus. Because of models such as these and others presented in this book, countries around the work instituted quarantines and social distancing to help slow the virus.

Another term used often on the news is that of herd immunity. Now let's briefly discuss the idea of herd immunity so we can understand what it implies.

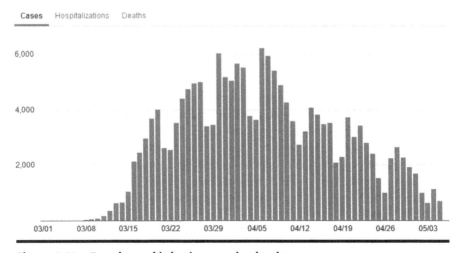

Figure 2.25   Bar chart of infection per day by date

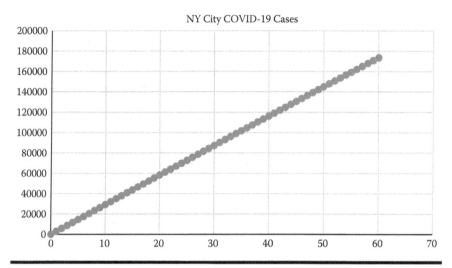

Figure 2.26   COVID-19 cases in New York City

It is essential to understand what our many scientists are referring to when they mention herd immunity.

2.5.1 *What Is Herd Immunity?*

When most of a population is immune to an infectious disease, this provides indirect protection—or herd immunity (also called herd protection)—to those who are not immune to the disease.

For example, if 80% of a population is immune to a virus, four out of every five people who encounter someone with the disease won't get sick (and won't spread the disease any further). In this way, the spread of infectious diseases is kept under control. Depending how contagious an infection is, usually 70% to 90% of a population needs immunity to achieve herd immunity. Many CDC officials as well as Dr. A. Fauci have briefed that 60% might allow us to achieve herd immunity. This is 60% of a given population.

How have we achieved herd immunity for other infectious diseases?

Measles, mumps, polio, and chickenpox are examples of infectious diseases that were once very common but are now rare in the U.S. because vaccines helped to establish herd immunity. We sometimes see outbreaks of vaccine-preventable diseases in communities with lower vaccine coverage because they don't have herd protection. (See the 2019 measles outbreak at Disneyland as an example.)

Let's revisit New York City for example. Herd immunity will occur when 0.60 of the 8.77 million have been infected. That is 5.262 million people. There are several very big assumptions here: (1) those that have had the disease cannot get it again and (2) they also cannot be carriers of the disease.

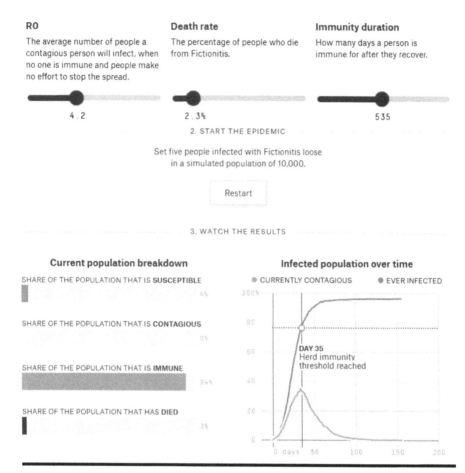

Figure 2.27 Interactive model concerning herd immunity (https://fivethirtyeight.com/features/without-a-vaccine-herd-immunity-wont-save-us/)

Unless the infection rates increase substantially, which they would as more people get infected, it will take years to get to 5.262 million. At this time, it is hard to estimate the rate. If there is a constant infectious rate of 338/day then we achieve herd immunity in New York City in one year.

There are two impacts of social distancing that we mention now. One is that good social distancing will slow the infectious rate and will flatten the curve. The second impact is slowing the curve slows the time until we reach herd immunity.

In an effort to show the fluidity of the COVID-19 analysis, an article entitled, "Without a vaccine, herd immunity won't save us," by Rodgers, Wolfe, and Bronner stated since measles took about a 93% infected rate before herd immunity happened, then perhaps COVID-19 may take a 95% infected rate.

Figure 2.21 show the results as a screenshot from the interactive program. If we apply to the entire USA with population 334,000,000 then about 317,300,000 will have to become infected. At the current death rate of 2% then about 6,346,000 Americans will die.

This is depicted in Figure 2.27.

In the next chapter, we couple the modeling effort with multiple interacting equations.

Chapter 3

Modeling Coupled Systems of Discrete Dynamical Systems

3.1 SIR and Other Models

In this section, we examine models of systems of difference equations (DDS). For selected initial conditions, we build numerical solutions to get a sense of long-term behavior for the system. For the systems that we will study, we will find their equilibrium values. We then explore starting values near the equilibrium values to see if by starting close to an equilibrium value, the system will:

a. remain close
b. approach the equilibrium value
c. not remain close.

What happens near these values gives great insight concerning the long-term behavior of the system. We can study the resulting pattern of the numerical solutions.

We start by illustrating some basic models before we address pandemic models. It is essential to attempt to understand how these models work

Example 3.1: Inventory System Analysis (Adapted from Meerschaert, 1999)

A store has a limited number of 20-gallon aquarium tanks. At the end of each week the manager inventories and places orders. Store policy is to order three new 20-gallon tanks at the end of each week if all the inventory has been sold.

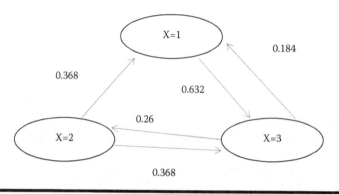

Figure 3.1 Flow diagram for inventory

If even one 20-gallon aquarium tank remains in stock, no new units are ordered that week. This policy is based upon the observation that the store sells on average one of the 20-gallon aquarium tanks each week. Is this policy adequate to guard against lost sales?

Historical data is used to compute the probabilities of demand given the number of aquarium tanks on hand. From this analysis, we obtain the following change diagram, Figure 3.1.

Problem Identification: Determine the number of 20-gallon aquariums to order each week.

Assumptions and variables: Let *n* represent the number of time periods in weeks. We define

$A (n)$ = the probability that one aquarium is in demand in week *n*.
$B (n)$ = the probability that two aquariums are in demand in week *n*.
$C (n)$ = the probability that three aquariums are in demand in week *n*.

We assume that no other incentives are given to the demand for aquariums over this time frame.

The Model:

The number of aquariums demanded in each time period is found using the paradigm, Future = Present + Change. Mathematically, this is written as:

$$A (n + 1) = A (n) - 0.632A (n) + 0.368\ B (n) + 0.184\ C(n)$$

$$B (n + 1) = B (n) - 0.632B (n) + 0.368\ C(n)$$

$$C (n + 1) = C (n) - 0.552\ C(n) + 0.632\ A(n) + 0.264\ B(n)$$

We seek to find the long term behavior of this system. We iterate the DDS

A	B	C
1	0	0
0.368	0	0.632
0.251712	0.232576	0.515712
0.273109	0.27537	0.451521
0.28492	0.267496	0.447584
0.285645	0.263149	0.451206
0.284978	0.262883	0.452139
0.284806	0.263128	0.452066
0.28482	0.263191	0.451989
0.284834	0.263186	0.45198
0.284836	0.263181	0.451983
0.284835	0.26318	0.451984
0.284835	0.263181	0.451984
0.284835	0.263181	0.451984
0.284835	0.263181	0.451984
0.284835	0.263181	0.451984
0.284835	0.263181	0.451984
0.284835	0.263181	0.451984
0.284835	0.263181	0.451984
0.284835	0.263181	0.451984
0.284835	0.263181	0.451984
0.284835	0.263181	0.451984

The long terms probabilities for aquarium demand are: $P (D = 1) = 0.284835$, $P (D = 2) = 0.263181$, and $P (D = 3) = 0.451984$, which can also be seen in Figure 3.2.

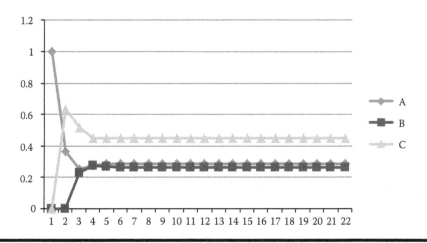

Figure 3.2   Plot of DDS for the aquarium inventory example

We iterate from near those equilibrium values and we find the sequences tend toward those values. We conclude the system has *stable* equilibrium values.

You should go back and change the initial conditions and see what behavior follows.

Interpretation: The long-term behavior shows that eventually (without other influences) the probabilities will be: *P (D = 1) = 0.284835, P (D = 2) = 0.263181,* and *P (D = 3) = 0.451984.*

We might want to try to attract new sales for aquariums by adding incentives for purchasing the aquariums.

Example 3.2: Competitive Hunter Models

Competitive hunter models involve species vying for the same resources (such as food or living space) in the habitat. The effect of the presence of a second species diminishes the growth rate of the first species. We now consider a specific example concerning trout and bass in a small pond. Hugh Ketum owns a small pond that he uses to stock fish and eventually allows fishing. He has decided to stock both bass and trout. The fish and game warden tells Hugh that after inspecting his pond for environmental conditions he has a solid pond for growth of his fish. In isolation, bass grow at a rate of 20% and trout at a rate of 30%. The warden tells Hugh that the interactions for the food affects trout more than bass. They estimate the interaction affecting bass is 0.0010 bass*trout and for trout is 0.0020 bass*trout. Assume no changes in the habitant occur.

Model:
Let's define the following variables

$B\ (n)$ = the number of bass in the pond after period n.
$T\ (n)$ = the number of trout in the pond after period n.
$B\ (n) * T\ (n)$ = interaction of the two species.

$$B\ (n + 1) = 1.20\ B(n) - 0.0010\ B(n) * T\ (n)$$

$$T\ (n + 1) = 1.30\ T\ (n) - 0.0020\ B(n) * T\ (n)$$

The equilibrium values can be found by allowing $X = B\ (n)$ and $Y = T\ (n)$ and solving for X and Y.

$$X = 1.2\ X - .001\ X * Y. \tag{3.1}$$

$$Y = 1.3\ Y - 0.0020\ X * Y \tag{3.2}$$

We rewrite equations (3.1) and (3.2) as

$$.2\ X - .001\ X * Y = 0 \tag{3.3}$$

$$.3\ Y - .002\ X * Y = 0 \tag{3.4}$$

We factor X out of (3.3) and Y out of (3.4) to obtain

$$X\,(.2 - .001\ Y) = 0$$

$$Y\,(.3 - .002\ X) = 0$$

Solving we find $X = 0$ or $Y = 2,000$ and $Y = 0$ or $X = 1,500$.

We want to know the long-term behavior of the system and the stability of the equilibrium points.

Hugh initially considers 151 bass and 199 trout for his pond. From Hugh's initial conditions, bass will grow without bound and trout will eventually die out. This is certainly not what Hugh had in mind.

Example 3.3: A Predator-Prey Model: *Foxes and Rabbits*

In the study of the dynamics of a single population, we typically take into consideration such factors as the "natural" growth rate and the "carrying capacity" of the environment. Mathematical ecology requires the study of populations that interact, thereby affecting each other's growth rates. In this module we study a very special case of such an interaction, in which there are exactly two species, one of which the predators eat the prey. Such pairs exist throughout nature:

- lions and gazelles,
- birds and insects,
- pandas and eucalyptus trees,
- Venus fly traps and flies.

To keep our model simple, we will make some assumptions that would be unrealistic in most of these predator-prey situations. Specifically, we will assume that

- the predator species is totally dependent on a single prey species as its only food supply,
- the prey species has an unlimited food supply, and
- there are no other threats to the prey other than the specific predator.

Vito Volterra (1860–1940) was a famous Italian mathematician who retired from a distinguished career in pure mathematics in the early 1920s. His son-in-law, Humberto D'Ancona, was a biologist who studied the populations of various species of fish in the Adriatic Sea. In 1926 D'Ancona completed a statistical study of the numbers of each species sold on the fish markets of three ports: Fiume, Trieste, and Venice. The percentages of predator species (sharks, skates, rays, etc.) in the Fiume catch are shown in Table 3.1.

We may assume that proportions within the "harvested" population reflect those in the total population. D'Ancona observed that the highest percentages of predators occurred during and just after World War I, when fishing was drastically curtailed. He concluded that the predator-prey balance was at its natural state during the war and that intense fishing before and after the war disturbed this natural balance – to the detriment of predators. Having no biological or ecological explanation for this phenomenon, D'Ancona asked Volterra if he could come up with a mathematical model that might explain what was going on. In a matter of months, Volterra developed a series of models for interactions of two or more species. The first and simplest of these models is the model that we will use for this scenario.

Alfred J. Lotka (1880–1949) was an American mathematical biologist (and later actuary) who formulated many of the same models as Volterra, independently and at about the same time. His primary example of a predator-prey system comprised a plant population and an herbivorous animal dependent on that plant for food.

Table 3.1 Percentages of Fiume Fish Catch

Percentages of Predators in the Fiume Fish Catch									
1914	1915	1916	1917	1918	1919	1920	1921	1922	1923
12	21	22	21	36	27	16	16	15	11

We repeat our two key assumptions:

- The predator species is totally dependent on the prey species as its only food supply.
- The prey species has an unlimited food supply and no threat to its growth other than the specific predator.

If there were no predators, the second assumption would imply that the prey species grows exponentially without bound, i.e., if $x = x\,(n)$ is the size of the prey population after a discrete time period n, then we would have x $(n + 1) = an\,x\,(n)$.

But there *are* predators, which must account for a negative component in the prey growth rate. Suppose we write $y = y\,(n)$ for the size of the predator population at time t. Here are the crucial assumptions for completing the model:

- The rate at which predators encounter prey is jointly proportional to the sizes of the two populations.
- A fixed proportion of encounters lead to the death of the prey.

These assumptions lead to the conclusion that the negative component of the prey growth rate is proportional to the product xy of the population sizes, i.e.,

$$x\,(n + 1) = x\,(n) + ax\,(n) - bx\,(n)\,y\,(n).$$

Now we consider the predator population. If there were no food supply, the population would die out at a rate proportional to its size, i.e., we would find $y\,(n + 1) = -cy\,(n)$.

We assume that is the simple case that the "natural growth rate" is a composite of birth and death rates, both presumably proportional to population size. In the absence of food, there is no energy supply to support the birth rate. But there is a food supply: the prey. And what's bad for hares is good for lynx. That is, the energy to support growth of the predator population is proportional to deaths of prey, so

$$y\,(n + 1) = y\,(n) - cy\,(n) + px\,(n)\,y\,(n).$$

This discussion leads to the discrete version of the Lotka-Volterra Predator-Prey Model:

$$
\begin{aligned}
x\,(n + 1) &= (1 + a)x\,(n) - bx\,(n)\,y\,(n) \\
y\,(n + 1) &= (1 - c)y\,(n) + px\,(n)\,y\,(n) \\
n &= 0, 1, 2, \ldots
\end{aligned}
\tag{3.5}
$$

where a, b, c, and p are positive constants.

The Lotka-Volterra model, equation 3.5, consists of a system of linked difference equations that cannot be separated from each other and that cannot be solved in closed form. Nevertheless, they can be solved numerically through iteration and graphed in order to obtain insights about the scenario being studied. In our foxes and hares scenario, let's assume this discrete model explained above. Further, data investigation yields the following estimates for the parameters $\{a, b, c, p\} = \{0.039, 0.0003, 0.12, \text{ and } 0.0001\}$. We plot the results in Figures 3.3–3.5.

If we ran this model for more iterations, we would find the plot of foxes versus rabbits spiral in a similar fashion as above. We conclude that the model appears reasonable. We could find the equilibrium values for the system. There are two sets of equilibrium points for rabbits and foxes (0, 0) and (1,200; 130). The orbits of the spiral indicate that the system is moving away from (1200, 130) so we conclude the system is not stable. In the exercise set, you will be asked to do more explorations.

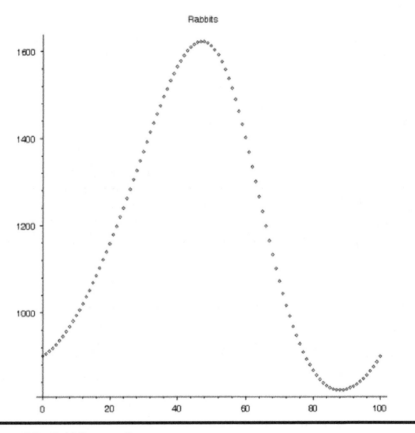

Figure 3.3 Rabbits over time

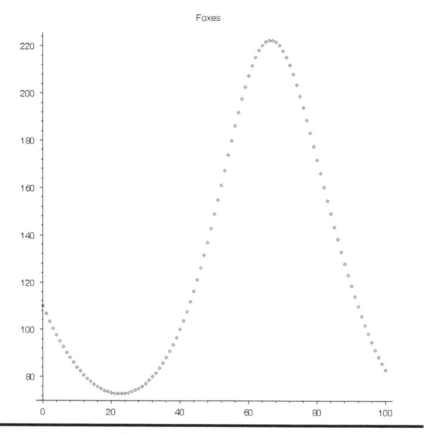

Figure 3.4 Foxes over time

Example 3.4: Discrete Basic **SIR Models** of Epidemics and Pandemics

Consider a disease that is spreading throughout the Unites States such as the new flu or COVID-19. The CDC is interested in knowing and experimenting with a model for this new disease prior to it actually becoming a "real" epidemic. Let us consider the population being divided into three categories: susceptible, infected, and removed. We make the following assumptions for our model:

- No one enters or leaves the community and there is no contact outside the community (we will relax this to see the impact later).
- Each person is either susceptible, S (able to catch this new flu); infected, I (currently has the flu and can spread the flu); or removed, R (already had the flu and will not get it again, and that includes death).
- Initially every person is either *S* or *I*.
- Once someone gets the flu or COVD-19 this year they cannot get it again. This is currently an actual assumption (not fact) by the CDC.

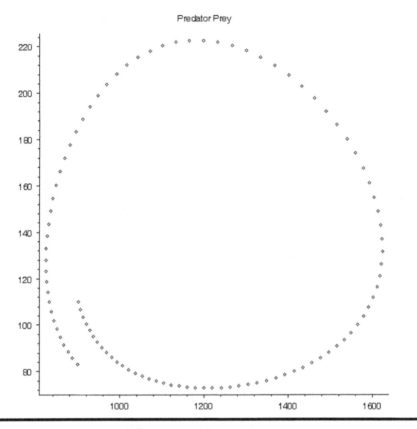

Figure 3.5 Foxes versus rabbits over time

- The average length of the disease is 2 weeks over which the person is deemed infected and can spread the disease.
- People can spread the disease while showing no symptoms.
- Our time period for the model will be per week.

The model we will consider is the SIR model (Allman, 2004).

Let's assume the following definition for our variables.

$S(n)$ = number in the population susceptible after period n.
$I(n)$ = number infected after period n.
$R(n)$ = number removed after period n.

Let's start our modeling process with $R(n)$. Our assumption for the length of time someone has the flu is 2 weeks. Thus, half the infected people will be removed each week,

$$R(n + 1) = R(n) + 0.5\ I(n)$$

The value, 0.5, is called the removal rate per week. It represents the proportion of the infected persons who are removed from infection each week. If real data is available, then we could do "data analysis" in order to obtain the removal rate.

$I(n)$ will have terms that both increase and decrease its amount over time. It is decreased by the number that are removed each week, $0.5*I(n)$. It is increased by the numbers of susceptible that come into contact with an infected person and catch the disease, $a\ S(n)\ I(n)$. We define the rate, a, as the rate in which the disease is spread or the transmission coefficient. We realize this is a probabilistic coefficient. We will assume, initially, that this rate is a constant value that can be found from initial conditions.

Let's illustrate as follows. Assume we have a population of 1,000 students in the dorms. Our nurse found 3 students reporting to the infirmary initially. The next week, 5 students came into the infirmary with flu-like symptoms. $I(0) = 3$, $S(0) = 997$. In week 1, the number of newly infected is 30.

$$5 = a\ I(n)\ S(n) = a\ (3) * (995)$$

$$a = 0.00167$$

Let's consider $S(n)$. This number is decreased only by the number that becomes infected. We may use the same rate, a, as before to obtain the model:

$$S(n + 1) = S(n) - a\ S(n)\ I(n)$$

Our coupled SIR model is

$$
\begin{aligned}
R(n + 1) &= R(n) + 0.5I(n) \\
I(n + 1) &= I(n) - 0.5I(n) + 0.00167I(n)S(n) \\
S(n + 1) &= S(n) - 0.00167S(n)I(n) \\
I(0) &= 3,\ S(0) = 997,\ R(0) = 0
\end{aligned}
\tag{3.6}
$$

The SIR Model, equation 3.6, can be solved iteratively and viewed graphically. Let's iterate the solution and obtain the graph to observe the behavior to obtain some insights; see Figures 3.6–3.9.

The worse of the flu or COVID-19 epidemic occurs around week 8, at the maximum of the infected graph. The maximum number is slightly larger than 400, from the table it is 427. After 25 weeks, slightly more than 9 people never get the flu or COVID-19.

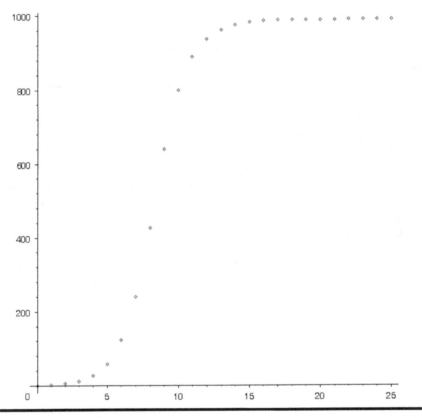

Figure 3.6 Plot of *R (n)* versus *n*

Example 3.5: Enhanced SIR Model for New York City

Let's illustrate as follows. Assume we have a population of 877,000,000 people in New York City. Our nurse found 3 people reporting to the hospital with symptoms initially. The next week, 125 people came to the hospital with flu-like symptoms. $I (0) = 3$, $S (0) = 87,699,997$. In week 1, the number of newly infected is 125.

$$125 = a\ I\ (n)\ S\ (n) = a\ (3) * (8769997)$$

$$a = 0.0000047565$$

Let's consider $S (n)$. This number is decreased only by the number that becomes infected. We may use the same rate, a, as before to obtain the model:

$$S\ (n + 1) = S\ (n) - aS\ (n)\ I\ (n)$$

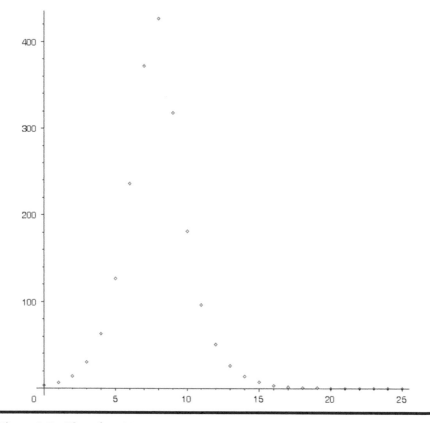

Figure 3.7   Plot of *m (n)* versus *n*

Our coupled SIR model is

$S(n + 1) = S(n) - 0.0000047565\ S(n)\ I(n)$

$I(n + 1) = I(n) - (1/3000)\ I(n) + 0.0000047565\ S(n)\ Ii(n)$

$R(n + 1) = R(n) + (1/3000)\ I(n)$

$R(0) = 0,\quad I(n) = 3,\quad S(n) = 87699997$

The SIR Model can be solved iteratively and viewed graphically. Let's iterate the solution and obtain the graph to observe the behavior to obtain some insights, see Figures 3.10.

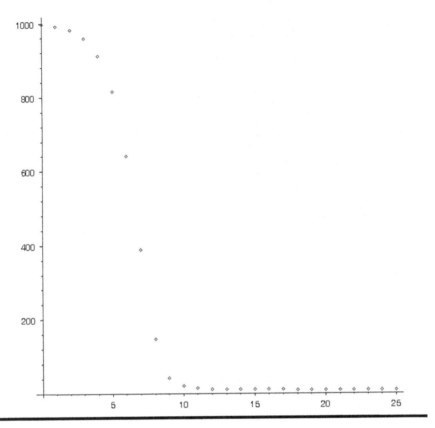

Figure 3.8   Plot of S (n) versus n

Insights show that in less than 100 days about 450,000 of the 8.77 million will have become infected.

The current death rate in New York City among those who are infected is about 7.8% so that we would expect 35,132 deaths at the end of 100 days.

Example 3.6: Enhanced SIR Model for the United States

We use the same model set up but different parameters. The US population is about 334 million.

Here is a graph of our output.

In about 70 months, we will be at about just above 40% of the entire population infected or about 136,000,000. If we examine with the current death rate of 5.79% (77,691/1,300,000 = 0.05973) then we would have had 8,123,280 deaths in the US.

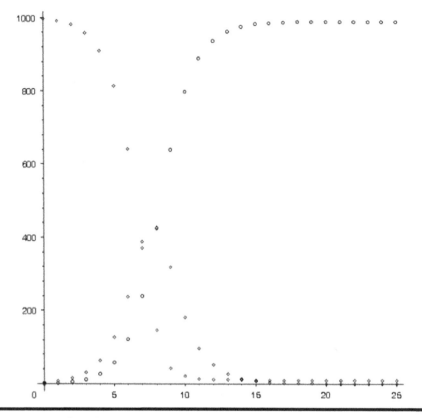

Figure 3.9 Plot of *R (n)*, *I (n)*, *S (n)* versus n together

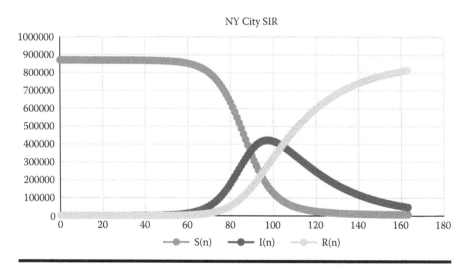

Figure 3.10 Plot of *SIR Model* for New York City versus *n in days*

So why are we not there? Well, we used quarantine and isolation as well as social distancing to affect the "real" results thus altering the parameters to our model and reducing the numbers of infected and ultimately the number of deaths.

Example 3.7: The SIRDS Model

Often modeling an epidemic or pandemic involves modeling deaths. Modeling deaths may or may not be a trivial endeavor. In the best modeling cases, deaths are a fraction (perhaps constant fraction) of the infected populations. Figure 3.11 illustrates this concept (Figure 3.12).

If this is the model that we use, we only need to apply the fractional deaths to the infections in the model. That is, the method we have used in our modeling. We can, in fact, decouple deaths since deaths only affect the total population.

An SIR model suggested by Bin Zao (see reference, Chapter 3) suggests a more complicated model. This might be in order for modeling COVID-19 as we still do not totally understand the virus.

This model has six classifications of variables.

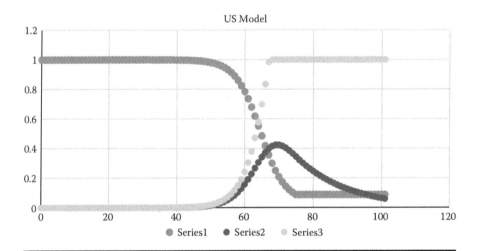

Figure 3.11 Plot of SIR for USA

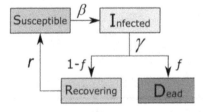

Figure 3.12 SIRD model

S (t) = Susceptible population who may become infected
E (t) = Infected by no symptoms
I (t) = Infected but not isolated or quarantined
Q (t) = Diagnosed and quarantined
D (t) = Potential victims
R (t) = Recovered from virus

The models may be viewed as follows

$$S\ (t+1) = S(t) + d_{qs}\,Q(t) - f(t) - d_{sq}S(t)$$

$$E\ (t+1) = E(t) + f(t) - eE(t) - d_{sq} - E(t)$$

$$D\ (t+1) = D(t) + d_{qd}\,Q(t) + di_d\,I(t) - (\varpi + \delta)\,D(t)$$

$$Q\ (t+1) = Q(t) + deq\ E(t) + dsq\ S(t) - dqs\ Q(t) - d_{qd}Q(t)$$

$$I\ (t+1) = I(t) + e\ E(t) - d_{id}I(t) - \delta I(t)$$

$$R(t+1) = R(t) + \varpi D\ (t)$$

Bin Zhao et al., estimated the parameters values as

$$d_{eq} = 4/35,\ \ d_{iq} = 1/3,\ \ d_{qt} = 1/3,\ \ d_{qs} = 0,\ \ \delta = 0.001,\ \ \varpi = 1/21,$$

$$e = 1/25$$

These parameters were based upon the following assumptions: incubation period is 7 days and average onset is 1 day after the incubation period. The average disease duration is 21 days. The mortality rate is estimated at about 2%. The daily conversion of suspected cases to confirmed cases is about 0.8 and newly admitted patients is about 0.2 of the confirmed cases per day.

3.2 Chapter Summary

In this chapter, we have explored the use of discrete dynamical systems using the paradigm, *Future = Present + Change*. We showed the importance of the change diagram to assist in the model equation formulation. We illustrated a solution process through iteration as well as visual approach. We introduced the concept of equilibrium and stable equilibrium to assist in analyzing the behavior of the DDS. We provided linear, non-linear, and systems of DDS examples.

Chapter 4

Modeling with Differential Equation

Here, we are going to first use a simple differential equation to look at the six-feet rule for social distancing. First, we state some CDC guidelines that are offered to reduce the spread of COVID-19.

According to the CDC, social distancing slows the spread of COVID-19.

Limiting face-to-face contact with others is the best way to reduce the spread of COVID-19.

4.1 What Is Social Distancing?

Social distancing, also called "physical distancing," means keeping space between yourself and other people outside of your home. To practice social or physical distancing:

- Stay at least six-feet away (about 2 arms' length) from other people.
- Do not gather in groups.
- Stay out of crowded places and avoid mass gatherings.

In addition to everyday steps to prevent COVID-19, keeping space between you and others is one of the best tools we have to avoid being exposed to this virus and slowing its spread locally and across the country and world.

Limit close contact with others outside your household in indoor and outdoor spaces. Since people can spread the virus before they know they are sick, it is important to stay away from others when possible, even if you – or they – have no symptoms. Social distancing is especially important for people who are at higher risk for severe illness from COVID-19.

Many people have personal circumstances or situations that present challenges with practicing social distancing to prevent the spread of COVID-19. Please see the following guidance for additional recommendations and considerations for:

- Households living in close quarters: how to protect those who are most vulnerable
- Living in shared housing
- People with disabilities
- People experiencing homelessness

4.1.1 Why Practice Social Distancing?

COVID-19 spreads mainly among people who are in close contact (within about six feet) for a prolonged period. Spread happens when an infected person coughs, sneezes, or talks, and droplets from their mouth or nose are launched into the air and land in the mouths or noses of people nearby. The droplets can also be inhaled into the lungs. Recent studies indicate that people who are infected but do not have symptoms likely also play a role in the spread of COVID-19.

It may be possible that a person can get COVID-19 by touching a surface or object that has the virus on it and then touching their own mouth, nose, or eyes. However, this is not thought to be the main way the virus spreads. COVID-19 can live for hours or days on a surface, depending on factors such as sunlight, humidity, and the type of surface. Social distancing helps limit opportunities to come in contact with contaminated surfaces and infected people outside the home.

Although the risk of severe illness may be different for everyone, anyone can get and spread COVID-19. Everyone has a role to play in slowing the spread and protecting themselves, their family, and their community.

4.1.2 Tips for Social Distancing

- Follow guidance from authorities where you live.
- If you need to shop for food or medicine at the grocery store or pharmacy, stay at least six feet away from others. Also consider other options:
  o Use mail-order for medications, if possible.
  o Consider a grocery delivery service.

- Cover your mouth and nose with a mask cloth face covering when around others, including when you have to go out in public, for example, to the grocery store.
  o Cloth face coverings should NOT be placed on children under age 2, anyone who has trouble breathing, or is unconscious, incapacitated, or otherwise unable to remove the mask without assistance.

o Keep at least six feet between yourself and others, even when you wear a face covering.

■ Avoid gatherings of any size outside your household, such as a friend's house, parks, restaurants, shops, or any other place. This advice applies to people of any age, including teens and younger adults. Children should not have in-person playdates while school is out. To help maintain social connections while social distancing, learn tips to keep children healthy while school's out.

■ Work from home when possible. See additional information for critical infrastructure workforce external icon from Cybersecurity and Infrastructure Security Agency (CISA).

■ Avoid using any kind of public transportation, ridesharing, or taxis, if possible.

■ If you are a student or parent, talk to your school about options for digital/distance learning.

Stay connected while staying away. It is very important to stay in touch with friends and family who don't live in your home. Call, video chat, or stay connected using social media. Everyone reacts differently to stressful situations and having to socially distance yourself from someone you love can be difficult. Read tips for stress and coping.

Before we model, here is an interesting article by Tina Saey, posted April 17, 2020.

In February, a man in Chicago brought food to and hugged two friends who had recently lost a family member. The next day, the man went to the funeral, where he comforted other mourners and shared a potluck meal. A few days later, he attended a family birthday party. The man had symptoms of a mild respiratory illness. Later he'd learn he had COVID-19.

His acts of condolence and celebration set off a chain reaction that sickened at least 16 people, three of whom died. At the time, social distancing measures weren't yet in place in Chicago. COVID-19 had yet to circulate widely in the area.

The case now serves as a cautionary tale, underscoring recommendations for people to keep their distance from anyone outside their immediate household, researchers report April 8 in *Morbidity and Mortality Weekly Report*. But how much distance is needed to avoid spreading the coronavirus?

Six feet (or two meters) has become the mantra. The World Health Organization and other experts have said SARS-CoV-2, the virus that causes COVID-19, is spread mainly by large droplets sprayed when people cough or sneeze, contaminating surfaces. So that degree of separation, combined with frequent hand-washing, was thought to be enough to halt or at least slow the spread of the virus.

But new evidence suggests six feet of distance may not be enough. If SARS-CoV-2 is airborne, as scientists think it may be, people could be infected simply by inhaling the virus in tiny aerosol droplets exhaled by someone talking or breathing.

What's actually safe is unknown. It may depend on many factors, including whether people are inside or outdoors, how loudly people are speaking, whether they are wearing masks, how well-ventilated a room is, and how far the virus can really fly.

Now, we need some data on speed in which a droplet or cough falls.

Solving an ODE

Define a simple ODE.

```
> ode := diff(y(t), t) =-6;
```

$$ode := \frac{d}{dt} y(t) = -6$$

Solve the ODE, ode.

```
> dsolve(ode)
```

$$y(t) = -6t + \_C1$$

Define initial conditions.

```
> ics := y(0) = 0
```

$$ics := y(0) = 0$$

Solve ode subject to the initial conditions ics.

```
> dsolve({ics, ode})
```

$$y(t) = -6t$$

```
> ode := diff(y(t), t) =-28;
```

$$ode := \frac{d}{dt} y(t) = -28$$

Solve the ODE, ode.

```
> dsolve(ode)
```

$$y(t) = -28t + \_C1$$

Define initial conditions.

```
> ics := y(0) = 0
```

$$ics := y(0) = 0$$

Solve ode subject to the initial conditions ics.

```
> dsolve({ics, ode})
```

$$y(t) = -28t$$

```
>
```

Figure 4.1 Screenshot from Pale for ODEs

In an article by Jan Dyer, how far a cough travels is discussed. Their research shows that as many as 3,000 droplets are exhaled with speeds between 6 and 28 m/s.

We will use these in our simple differential equations model. We will build two models: one based on 6 m/s and the other based on 28 m/s.

Assuming we start at v (0) = 0 then our two models are shown in the Maple screenshot Figure 4.1

We have $-3t^2$ and $-14\ t^2$ as our distance model where negative means they are falling. If the average person is 5' 5" or 65", we can convert to meters where 1 foot = 0.3048 m.

So 5.42 feet is 1.651 meters.

If you assume people walk at 1 m/s and we have the falling droplets at $-6t$ or $-28\ t$ m/s, then after 2 sec (which is about six feet for the person walking) the droplets would have fallen almost to the ground.

Six feet might be the minimum distance required.

4.2 How Far Does a Cough Travel?

Researchers reveal the trajectory of a cough and the role it plays in spreading viral infections like influenza. We present some information modeled by Jan Dyer on January 27, 2017:

> "A single cough can propel as many as 3,000 droplets into the air at a velocity of 6 to 28 m/s. The droplets travel in what classic fluid mechanics calls a '2-stage jet': the starting jet (when the cough starts) and interrupted jet (when the cough stops). After the original cough ends, a 'leading vortex' carries particles forward, but as the momentum slows, particles fall out of the jet according to researchers from the University of Hong Kong and Shenzhen Institute of Research and Innovation, both in China, who studied cough trajectories and the implications for disease transmission in buildings."

> "Once the penetration velocity drops below 0.01 m/s, environmental factors, such as ventilation and human body temperature, begin to influence the flow. Beyond 1 to 2 m, the exhaled air stream dissolves into the room airflow, and the pathogen-containing droplets or droplet nuclei are dispersed according to the global airflow in the room."

Let's model this:

$dv/dt = -17$

$v = -17\ dt$

Assuming v (0) = 0 then

$v = -17\ t$

assuming $d(0) = 1.828$ m (6 feet)

Distance $d(t) = -8.5\ t^2 + 1.818\ t$

Walking at normal speed, by the time we go six feet the droplets will have almost all settled onto the floor. Our mouth, nose, and eyes should be safe from the droplets entering them. However, if we are running or biking, so that our speed of movement is greater, the droplets will most likely be in our path and we are subject to being contaminated with the virus.

Example 4.1: Drug Dosage as a DE

In Maple, we solve the ODE, $d(ct)/dt = -.3\ c(t)$ with initial condition $c(0) = 1$.

The solution is $c(t) = e^{-.3t}$

Figure 4.2 illustrates this ODE solution.

We assume that the first dose is administered at time $t = 0$, and the concentration of the drug in the bloodstream before then is zero. At $t = 0$ the concentration jumps to b, and then for $0 < t < h$, the concentration decays according to (1). At $t = h$, another dose is administered, and the concentration increases by b.

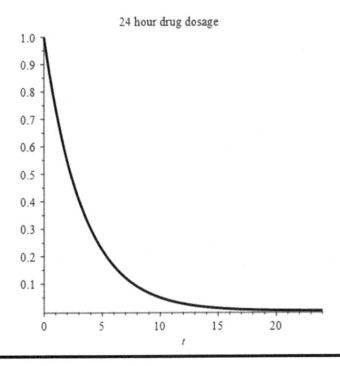

Figure 4.2   Plot of our ODE for drug dosage

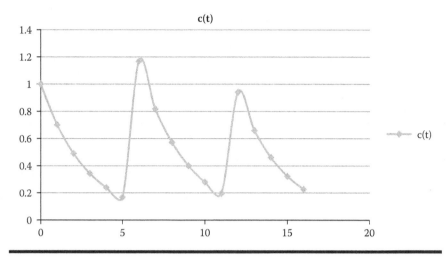

Figure 4.3 Plot of concentrations of c(t) for a drug administered periodically. In this example, *r* = –0.3, *h* = 6, and *b* = 1. At *t* = 0, h, 2h, ..., *c(t)* increases by *b*; otherwise the concentration decays according to the equation.

The concentration then decays for $h < t < 2\,h$, and the process continues. We expect the plot of the concentration to look like the graph shown in Figure 4.3.

According to Weckesser (Colgate University), the equilibrium is the limit as x → infinity. The equation from our simple model is,

$$\frac{be^{-rh}}{(1 - e^{-rh})}$$

For our model b = 1, h is every 2 hours, r = 0.3. For more information, see (http://math.colgate.edu/~wweckesser/math312Spring05/handouts/PeriodicDrugDose.pdf).

The equilibrium value is 1.2164. We assume that this amount of drug is what is required to treat the disease (see Giordano, Fox, and Horton for more information).

Chapter 5

Systems of Differential Equations

5.1 Model of Systems

Interactive situations occur in the study of economics, ecology, electrical engineering, mechanical systems, control systems, systems engineering, and so forth. For example, the dynamics of population growth of various species is an important ecological application of applied mathematics. The spread of a disease, especially for a pandemic, is critical for analysis.

Example 5.1: Continuous SIR Models of Epidemics

Consider a disease that is spreading throughout the Unites States such as the new flu. The CDC is interested in knowing and experimenting with a model for this new disease before it actually becomes a "real" epidemic. Let us consider the population is divided into three categories: susceptible, infected, and removed. We make the following assumptions for our model:

- No one enters or leaves the community and there is no contact outside the community.
- Each person is either susceptible, S (able to catch this new flu); infected, I (currently has the flu and can spread the flu); or removed, R (already had the flu and will not get it again, which includes death).
- Initially every person is either S or I.
- Once someone gets the flu this year, they cannot get it again.
- The average length of the disease is two weeks over which the person is deemed infected and can spread the disease.
- Our time period for the model will be per week.

The model we will consider is the SIR model (Allman, 2004). Let's assume the following definitions for our variables:

$S(n)$ = number in the population susceptible after period n.
$I(n)$ = number infected after period n.
$R(n)$ = number removed after period n.

Let's start our modeling process with $R(n)$. Our assumption for the length of time someone has the flu is two weeks. Thus, half the infected people will be removed each week,

$$\frac{dR}{dt} = 0.5 * I(t)$$

The value, 0.5, is called the removal rate per week. It represents the proportion of the infected persons who are removed from infection each week. If real data is available, then we could do "data analysis" in order to obtain the removal rate.

$I(t)$ will have terms that both increase and decrease its amount over time. It is decreased by the number that are removed each week, $0.5 * I(t)$. It is increased by the number of susceptibles who come in contact with an infected person and catch the disease, $a\,S(t)\,I(t)$. We define the rate, a, as the rate in which the disease is spread or the transmission coefficient. We realize that this is a probabilistic coefficient. We will assume, initially, that this rate is a constant value that can be found from initial conditions.

Let's illustrate as follows. Assume we have a population of 1,000 students in the dorms. Our nurse found three students reporting to the infirmary initially. The next week, five students came into the infirmary with flu-like symptoms. $I(0) = 3$, $S(0) = 997$. In week 1, the number of newly infected is 30.

$$5 = a\,I(n)\,S(n) = a(3) * (995)$$

$$a = 0.00167.$$

Let's consider $S(t)$. This number is decreased only by the number that becomes infected. We may use the same rate, a, as before to obtain the model,

$$\frac{dS}{dt} = -0.00167 \cdot S(t) \cdot I(t)$$

Our coupled SIR model is shown in the systems of differential equations (equation 5.1) as,

$$\frac{dR}{dt} = 0.5I(t)$$

$$\frac{dI}{dt} = -0.5I(t) + 0.00167I(t)S(t)$$ (5.1)

$$\frac{dS}{dt} = -0.00167S(t)I(t)$$

$$I(0) = 3, \ S(0) = 997, \ R(0) = 0$$

The SIR model given earlier can be solved iteratively and viewed graphically. Let's iterate the solution and obtain the graph to observe the behavior to obtain some insights.

We use technology to assist in the solution but the specific technology is not important (Figures 5.1–5.8).

> *removed* := *diff* (*r* (*t*), *t*) = 0.5 · *Inf* (*t*);

$$removed := \frac{d}{dt}r(t) = 0.5 \ Inf(t)$$

> *infect* := *diff* (*Inf* (*t*), *t*) = −0.5 · *Inf* (*t*) + 0.00167 · *s* (*t*) · *Inf* (*t*);

$$infect := \frac{d}{dt}Inf(t) = -0.5 \ Inf(t) + 0.00167 \ s(t) \ Inf(t)$$

> *succept* := *diff* (*s* (*t*), *t*) = −0.00167 · *s* (*t*) · *Inf* (*t*);

$$succept := \frac{d}{dt}s(t) = -0.00167 \ s(t) \ Inf(t)$$

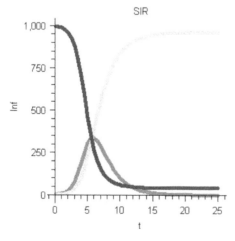

Figure 5.1   SIR graphical model for Example 5.1

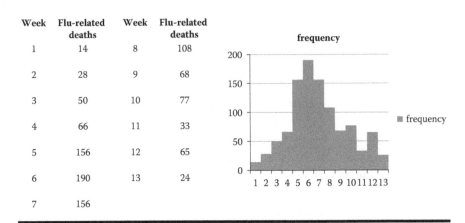

Week	Flu-related deaths	Week	Flu-related deaths
1	14	8	108
2	28	9	68
3	50	10	77
4	66	11	33
5	156	12	65
6	190	13	24
7	156		

Figure 5.2 Data graph for histogram for Hong Kong flu

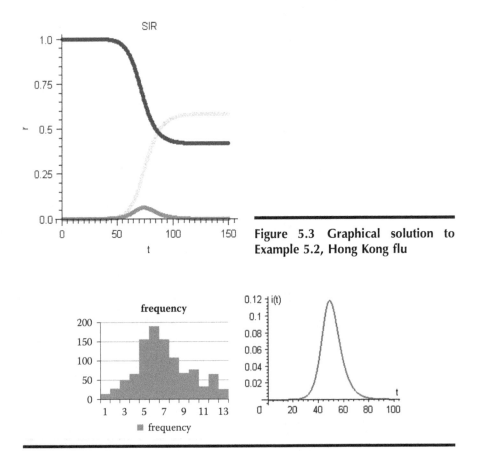

Figure 5.3 Graphical solution to Example 5.2, Hong Kong flu

Figure 5.4 Graphical comparison of the actual data for the Hong Kong flu or our model's graphical curve

APRIL 5, 1918

Weekly public health report tells of first U.S. flu fatalities with three deaths in Haskell, Kan.

100,000 TO 195,000

Number of U.S. deaths in October 1918 alone during the deadly second wave of the pandemic, which hit from September through November, beginning in the Boston area

675,000

Number of deaths in the U.S. attributed to the pandemic

40%

Decline in shipyard productivity reported in New York City due to flu illnesses in the midst of World War I

12 YEARS

The number of years that life expectancy in the U.S. had fallen by 1919 because of the pandemic, to 36.6 years for men and 42.2 years for women

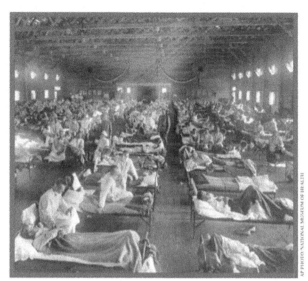

Influenza victims crowd into an emergency hospital near Fort Riley, Kan., in this 1918 file photo. The Spanish flu pandemic killed millions worldwide and officials say if the next pandemic resembles the birdlike 1918 flu, up to 1.9 million Americans could die.

500 MILLION

The number of people worldwide who became infected with the virus, about one-third of the world's population

50 MILLION

Number of deaths worldwide attributed to the pandemic

ZERO

Number of laboratory tests to diagnose, detect or characterize the flu in 1918

2005

The year CDC researchers were able to physically reconstruct the 1918 pandemic virus with reverse genetics. They found the HA and PB1 virus genes made for "maximum replication and virulence"

– Centers for Disease Control and Prevention; numbers for deaths and infections are only estimates but are based on CDC research.

Figure 5.5   Clipping from paper with data available

> *Model1* := {*removed*, *infect*, *succept*};

$$Model1 := \left\{ \frac{d}{dt}Inf(t) = -0.5\ Inf(t) + 0.00167\ s(t)\ Inf(t), \right.$$

$$\left. \frac{d}{dt}r(t) = 0.5\ Inf(t),\ \frac{d}{dt}s(t) = -0.00167\ s(t)\ Inf(t) \right\}$$

> *vars* := {*r*(*t*), *Inf*(*t*), *s*(*t*)};

$$vars := \{s(t),\ Inf(t),\ r(t)\}$$

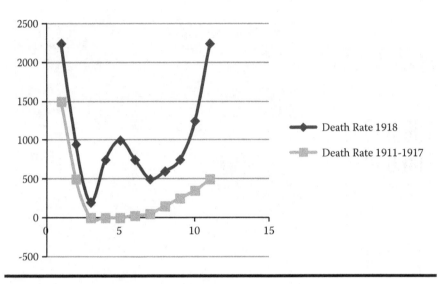

Figure 5.6 *U* and *W* **curves**

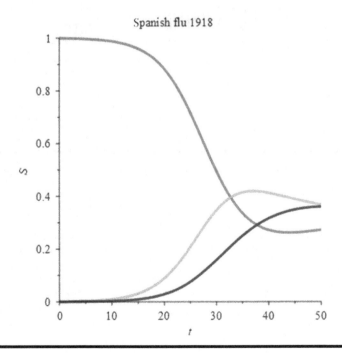

Figure 5.7 **Building the curves**

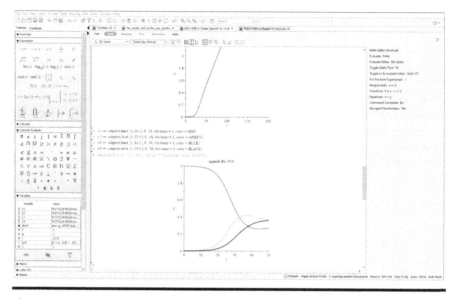

Figure 5.8   Screenshots of SIR models

> $init1 := [s(0) = 997, r(0) = 0, Inf(0) = 3];$

$init1 := [s(0) = 997, r(0) = 0, Inf(0) = 3]$

> $domain := t = 0 ..25;$

$domain := t = 0 ..25$

> $L := DEplot(Model1, vars, domain, [init1], stepsize = 0.5,$
$scene = [t, s], arrows = NONE, linecolor = blue):$

> $H := DEplot(Model1, vars, domain, [init1], stepsize = 0.5,$
$scene = [t, Inf], arrows = NONE, linecolor = red):$

> $G := DEplot(Model1, vars, domain, [init1], stepsize = 0.5,$
$scene = [t, r], arrows = NONE):$

> $display(\{L, H, G\}, title = 'SIR');$

In this example, we see that the maximum number of infected persons occurs at about day 7. Everyone survives and not everyone gets the flu. Let's see what happens in another case example.

Example 5.2: Hong Kong Flu (From Lang and Moore, MAA)

During the winter of 1968–1969, the United States was swept by a virulent new strain of influenza, named *Hong Kong flu* for its place of discovery. At that time, no flu vaccine was available, and therefore, many more people were infected than would be the case today. We will study the spread of the disease through a single urban population, that of New York City. The data displayed in the following table are weekly totals of "excess" pneumonia-influenza deaths, that is, the numbers of such deaths in excess of the average numbers to be expected from other sources. The graph (Figure 5.2) displays the same data (source: Centers for Disease Control).

Relatively few flu sufferers die from the disease or its complications, even without a vaccine. However, we may reasonably assume that the number of excess deaths in a week was proportional to the number of new cases of flu in some earlier week, say, three weeks earlier. Thus, the figures in the table reflect (proportionally) the rise and subsequent decline in the number of new cases of Hong Kong flu. We will model the spread of such a disease so that we can predict what might happen with similar epidemics in the future.

At any given time during a flu epidemic, we want to know the number of people who are infected. We also want to know the number who have been infected and have recovered, because these people now have an immunity to the disease. (As a matter of convenience, we include in the recovered group the relative handful who do not recover but die – they too can no longer contract the disease.) If we ignore movement into and out of the infected area, then the remainder of the population is still susceptible to the disease. Thus, at any time, the fixed total population (approximately 7,900,000 in the case of New York City in the late 1960s) may be divided into three distinct groups:

The first set of dependent variables counts *people* in each of the groups, each as a function of time:

$S = S(t)$	is the number of *susceptible* individuals,
$I = I(t)$	is the number of *infected* individuals, and
$R = R(t)$	is the number of *recovered* individuals.

The second set of dependent variables represents the *fraction* of the total population in each of the three categories. So, if N is the total population (7,900,000 in our example), we have:

$s(t) = S(t)/N$	the susceptible fraction of the population,
$i(t) = I(t)/N$	the infected fraction of the population, and
$r(t) = R(t)/N$	the recovered fraction of the population.

It may seem more natural to work with population counts, but some of our calculations will be simpler if we use the fractions instead. The two sets of dependent variables are proportional to each other, so either set will give us the same information about the progress of the epidemic.

Finally, we complete our model by giving each differential equation an initial condition. For this particular virus – Hong Kong flu in New York City in the late 1960s – hardly anyone was immune at the beginning of the epidemic, so almost everyone was susceptible. We will assume that there was a trace level of infection in the population, say, 10 people. Thus, our initial values for the population variables are:

$S(0) = 7,900,000$
$I(0) = 10$
$R(0) = 0$

In terms of the scaled variables, these initial conditions are:

$s(0) = 1$
$i(0) = 1.27 \times 10^{-6}$
$r(0) = 0$

$$\frac{ds}{dt} = -b\,s(t)\,i(t), \qquad s(0) = 1,$$

$$\frac{di}{dt} = b\,s(t)\,i(t) - k\,i(t), \qquad i(0) = 1.27 \times 10^{-6},$$

$$\frac{dr}{dt} = k\,i(t), \qquad r(0) = 0.$$

This model is identical to the one we just developed. So what is different? We must keep in mind the number lost due to deaths from the flu. We have some data so we can estimate the parameters we need. We don't know values for the parameters b

and *k* yet, but we can estimate them, and then adjust them as necessary to fit the excess death data. We have already estimated the average period of infectiousness at three days, so that would suggest k = 1/3. If we guess that each infected would make a possibly infecting contact every two days, then b would be 1/2. We emphasize that this is just a guess. The following plot shows the solution curves for these choices of *b* and *k*.

> *removed := diff (r (t), t) = 0.33333333 · Inf (t);*

$$removed := \frac{d}{dt}r(t) = 0.33333333 \; Inf(t)$$

> *infect := diff (Inf (t), t) = −0.333333 · Inf (t) + 0.5 · s (t) · Inf (t);*

$$infect := \frac{d}{dt}Inf(t) = -0.333333 \; Inf(t) + 0.5 \; s(t) \; Inf(t)$$

> *succept := diff (s (t), t) = −0.5 · s (t) · Inf (t);*

$$succept := \frac{d}{dt}s(t) = -0.5 \; s(t) \; Inf(t)$$

> *Model1 := {removed, infect, succept};*

$$Model1 := \left\{ \frac{d}{dt}Inf(t) = -0.333333 \; Inf(t) + 0.5 \; s(t) \; Inf(t), \right.$$

$$\left. \frac{d}{dt}s(t) = -0.5 \; s(t) \; Inf(t), \frac{d}{dt}r(t) = -0.33333333 \; Inf(t) \right\}$$

> *vars := {r (t), Inf (t), s (t)};*

$$vars := \{s(t), Inf(t), r(t)\}$$

> *init1 := [s (0) = 1, r (0) = 0, Inf (0) = 1.27 · 10^{−6}];*

$$init1 := [s(0) = 1, r(0) = 0, Inf(0) = 0.000001270000000]$$

> *domain* := *t* = 0 ..150;

domain := *t* = 0 ..150

> *L* := *DEplot* (*Model1, vars, domain,* [*init1*], *stepsize* = 0.5,
   scene = [*t, s*], *arrows* = *NONE, linecolor* = *blue*):

> *H* := *DEplot* (*Model1, vars, domain,* [*init1*], *stepsize* = 0.5,
   scene = [*t, Inf*], *arrows* = *NONE, linecolor* = *red*):

> *G* := *DEplot* (*Model1, vars, domain,* [*init1*], *stepsize* = 0.5,
   scene = [*t, r*], *arrows* = *NONE*):

> *display* ({*L, H, G*}, *title* = '*SIR*');

Let's interpret these results. The model shows the number of susceptible decreasing as a percent from 100% to just less than 50%. We also see the number removed goes from 0% to close to 60%. Finally, we see the number infected peaking around 75 to a value of about 5%. We started with 7,900,000 people. So, about 395,000 have the flu at the maximum time period. The area under the curve would represent the total number that had been infected in the epidemic. And, what about our losses due to deaths? The average deaths per week was 79.6 or about 0.001% of the population.

   Now let's compare our model with the data. Recall that these were the numbers of deaths each week that could be attributed to the flu epidemic. If we assume that the fraction of deaths among infected individuals is constant, then the number of deaths per week should be roughly proportional to the number of infected people in some earlier week. We repeat the graph of the data, along with the graph of i (**t**) with k = **1/3** and b = **6/10** and we examine the graphs. The shape of the model appears reasonable.

Example 5.3: 1918 Spanish Flu

Over 100 years ago, the "mother of all pandemics" was sweeping the world. The flu pandemic was caused by an airborne H1N1 avian virus and killed an estimated 1% to 2% of the world's population in 1918 and 1919. The victims were primarily young and often healthy adults. The pandemic struck during World War I, killing more than the 17 million who died in that conflict; it is still considered one of the deadliest disease outbreaks in recorded history. The 1918 pandemic came to be dubbed the Spanish flu not because it originated there but because Spain was neutral in the war and so reported freely on the outbreak. The United States and other countries at war suppressed information about the severity of the disease so as not to damage morale.

Let's model using the SIR model.

The facts were that the world population was 1.8 billion, 50 million became infected, and over 17.4 million people died. With these facts, we can create some parameters to use.

```
> 
> with(DETools) : with(plots) :
> r := 0.3333 : v := 0.1 : g := 0.1 : f := 0.5 :
> dsn4 := dsolve( { d/dt S(t) =-r·S(t)·II(t) + v·R(t), d/dt II(t) = r·S(t)·II(t) - g·II(t), d/dt R(t) = g·II(t) - v·R(t), diff(N(t), t) = g·f·II(t), N(0) = 0, S(0) = 1, II(
    = .001, R(0) = 0}, numeric, range = 0..100);
                                    dsn4 := proc(x_rkf45) ... end proc
> sol1 := seq(dsn4(t), t = 0..200) :
```

Note the shapes of the following curves.

The U and W shapes are for the flu- and pneumonia-related deaths. The significance is the resurgence over time.

Building the curves came from additional models and curves.

We mimic these in our interpetation of the models.

Note: RK-4 was used to numerically solve this system of differential equations.

Example 5.4: SIR Model Applied to New York City for COVID-19

Let's illustrate as follows. Assume we have a population of 877,000,000 people in New York City. Our nurse found three people reporting to the hospital with symptoms initially. The next week, 125 people came to the hospital with flu-like symptoms. *I (0) = 3, S (0) = 87,699,997.* Now, the number infected is 1.3 million.

$$1.3 = a\, I(n) S(n) = a(8.77m)$$

$$a = 0.1482$$

Let's consider *S (n).* This number is decreased only by the number that becomes infected. We may use the same rate, *a,* as before to obtain the model,

Our coupled SIR model is,

$$dS/dt = -0.482\, S(t) I(t)$$

$$dI/dt = -0.03 \ I\,(t) \ + \ 0.1482 \ S\,(t) \ Ii \ ()$$

$$dR/dt = 0.03I \ (t)$$

$$R\,(0) = 0, \ I(0) = 3, \ S\,(0) = 87699997.$$

We scaled the entries so that the number of susceptible is 100% or 1.

The SIR model can be solved iteratively and viewed graphically as an ODE. Let's iterate the solution and obtain the graph to observe the behavior to obtain some insights, see Figure 5.9.

Based upon this model, the number infected after 150 days will be about 50% of the population and the deaths based upon the death rate would be about 0.02 ∗ .5 about 1% or 87,700. Again, this is the worst case model and since Gov. Cuomo instituted his policies of social distancing and closing certain businesses, he has slowed the infection rate substantially.

Example 5.5: SIR Model Template with Slider Scale

In an SIR model with a MAPLE template available at www.maplesoft.com/ application, we modify some parameters by using the slider bar in Figure 5.10 and obtain some interesting curves for *S (n)*, *I (n)*, and *R (n)*.

These oscillating curves are similar to one in a *New York Times* article on May 8. We used this to mimic what we saw in the *New York Times* article.

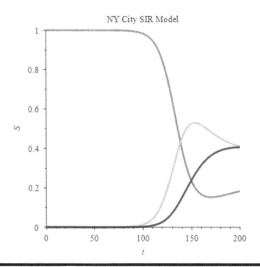

Figure 5.9   SIR model for NY city

▼ Interactive SIR model

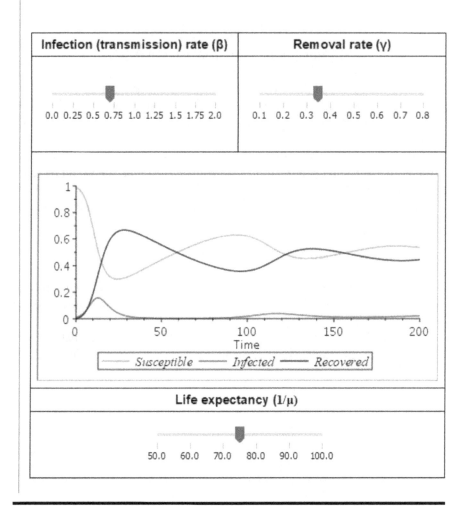

Figure 5.10 Screenshot from available SIR model in Maple

We use the SIR model with births and deaths by G. Edenharter, Edenharter Research (09/06/2015) available on line.

The model summary shows curves very similar to the prediction shown in the articles. Notice that it shows waves of progression based upon opening the economies.

The parameters are shown in Figure 5.11.

Model summary		
Parameter name	**Symbol**	**Value**
Susceptibles, fraction at t=0	$S(0)$	0.99000
Infected, fraction at t=0	$I(0)$	0.01000
Recovered, fraction at t=0	$R(0)$	0.00000
Birth rate=Death rate	μ	0.01333
Life expectancy	$\dfrac{1}{\mu}$	75.00000
Infection (transmission) rate	β	0.70051
Removal rate	γ	0.34643
Basic reproduction ratio	$R_0 = \dfrac{\beta}{\gamma + \mu}$	1.94714
S_{eq}	$\dfrac{1}{R_0}$	0.51357
I_{eq}	$\dfrac{\mu}{\beta} \cdot (R_0 - 1)$	0.01803
R_{eq}	$1 - S_{eq} - I_{eq}$	0.46840

Figure 5.11 Screenshot of parameters in the current model

5.2 Chapter Summary

No matter what model is used, we find the results are pretty consistent. The models here are merely single value predictors and not interval predictors. We also point out that the parameters used were based on the most current situations where social distancing and quarantines are in place.

Chapter 6

Probabilistic Models

6.1 Introduction

Dealing with probabilistic models means we are dealing with uncertainty. This is certainly true these days while discussing COVID-19. Because almost all medical research arenas are still unsure about COVID-19, that makes modeling more difficult. In this chapter, we pose some probabilistic models for consideration. In this chapter, we examine empirical models based upon collected data, applications from a few selected distributions, and the use of regression models for explaining and predicting the data.

6.2 Empirical Model and Forecasts

We use data from the COVID-19 http://www.worldometers.info/coronavirus/country (Figures 6.1 and 6.2).

With empirical models, we need to do some data manipulations and computations.

We can use this cumulative distribution function to find probabilities.

Let's look at the possible impact of social distancing. We built two DDS models – one with exponential growth based upon contacts and spread, and the other reduced model based upon Notre Dames' network model for social distancing which should reduce the reduction rate.

Notre Dames' deterministic model shows that one person can infect up to 39 people, as shown in Figure 6.3.

If we apply probabilities attributes, then the model shows the decease is upto eight persons infected by one person, as shown in Figure 6.4.

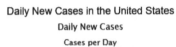

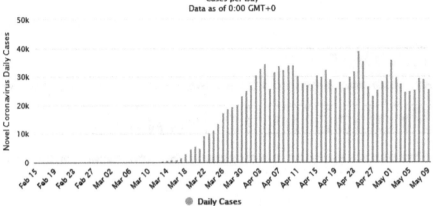

Figure 6.1   Data for daily new cases in the United States

Intervals	Counts	PDF	CDF		
10	55	0.00184	0.001842546	10	0.001843
20	300	0.01005	0.011892797	20	0.011893
30	1795	0.06013	0.072026801	30	0.072027
40	6240	0.20905	0.281072027	40	0.281072
50	7785	0.2608	0.541876047	50	0.541876
60	5646	0.18915	0.731021776	60	0.731022
70	4940	0.16549	0.896515913	70	0.896516
80	3090	0.10352	1.000033501	80	1.000034

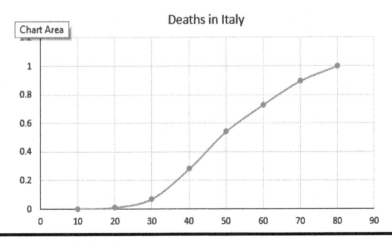

Figure 6.2   Deaths in Italy

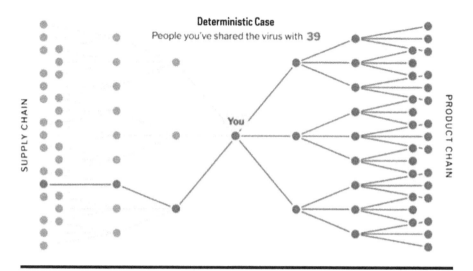

Figure 6.3   Notre Dame Network deterministic infection model snapshot

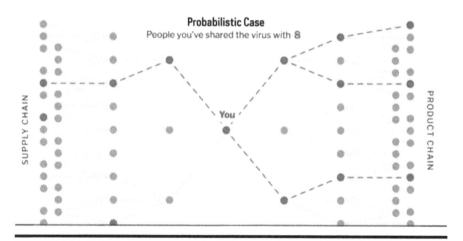

Figure 6.4   Notre Dame Network probabilistic infection model snapshot

They apply social distancing with the probabilistic attributes and find that there is a good chance that one infected person does not infect any others (Figure 6.5).

Further, if we use some results, we can build a DDS with some probabilistic components and iterate for 14 days.

	Social Distancing Model	
n	i(n)	IS(n)
0	2	2
1	3	2.333333
2	4.5	2.722222
3	6.75	3.175926
4	10.125	3.705247
5	15.1875	4.322788
6	22.78125	5.043253
7	34.17188	5.883795
8	51.25781	6.864427
9	76.88672	8.008499
10	115.3301	9.343248
11	172.9951	10.90046
12	259.4927	12.7172
13	389.239	14.83673
14	583.8585	17.30952

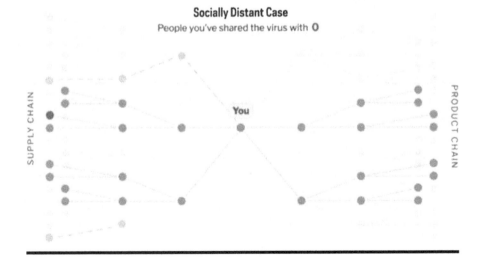

Socially Distant Case

People you've shared the virus with ○

Figure 6.5 Notre Dame Network social distance infection model snapshot

After two weeks, the number infected by two initially infected people is only 18 (rounded up) who have been under social distancing but would have been 584 infected if social distancing was not maintained. The results is a decrease by 97% of those infected. Results show that no matter what model is used, social distancing works to reduce the spread of the virus.

6.3 Markov Chain Models

A Markov chain is a discrete-time stochastic model. It is an appropriate generalization of the discrete time discrete dynamical systems presented in Chapters 2 and 3. Although the models can range from simple to very complex, there are many applications. Here, we introduce a simple Markov chain for the spread of COVID-19 to see how it works.

Markov chains begin, like a DDS, with a change diagram, now called a state transition diagram.

Let's start with an example for COVID-19 with only four transitions – susceptible, infected, recovered, and dead. We go back to the literature to obtain the transition probabilities.

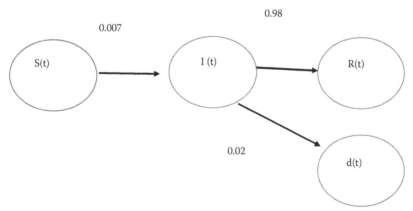

Transition state diagram:

We assume that once a patient recovers, they are immune and do not return to the susceptible population. We ran this model until 95% of the initial population has become infected. This is one value related to herd immunity. In our model, it take 441 days to achieve herd immunity. At that point, about 2% of the infected population (0.95 ∗ 334 M) have died. The number estimated to die from COVID-19 is about 6.4 million.

Other researchers used similar Markov chain models to show the effectiveness of social distancing (see Cano and Morales, 2020) based upon a population of 60 million.

We present a summary table of some of their results.

Social Distancing	Percentage Dead	Number Dead	Day Resolved
Almost perfectly executed	0.04%	21,474	90
50% effective	0.13%	79,781	250
Moderate, 75% effective	0.09%	32,998	189
Almost none	0.55%	330,094	112

6.4 Regression Models and Regression Analysis

Why do we use regression models? We might use them to help explain what we see in the data or we might use the model to help predict future outcomes. We start with basic least squares and a brief description of some diagnostics.

6.4.1 Curve Fitting Criterion

We will briefly discuss our curve fitting criterion: least squares or linear regression. The method of least-squares curve fitting, also known as **ordinary least squares** and **linear regression**, is simply the solution to a model that minimizes the sum of the squares of the deviations between the observations and predictions. Least squares will find the parameters of the function, $f(x)$ that will

$$\text{Minimize } S = \sum_{j=1}^{m} \left[y_i - f(x_j) \right]^2. \tag{6.1}$$

For example, to fit a proposed model $y = kx^2$ to a set of data, the least-squares criterion requires the minimization of equation 6.2. Note in equation 6.2, k is a slope.

$$\text{Minimize } S = \sum_{j=1}^{5} \left[y_i - kx_j^2 \right]^2 \tag{6.2}$$

Minimizing equation 6.2 is achieved using the first derivative, setting it equal to zero, and solving for the unknown parameter, k.

$$\frac{ds}{dk} = -2 \sum x_j^2 (y_j - kx_j^2) = 0. \text{ Solving for k: } k = \left(\sum x_j^2 y_j \right) \Big/ \left(\sum x_j^4 \right). \tag{6.3}$$

Table 6.1   Data Used

X	0.5	1.0	1.5	2.0	2.5
Y	0.7	3.4	7.2	12.4	20.1

Given the data set in Table 6.1, we will find the least squares fit to the model, $y = kx^2$.

Solving for k: $k = (\sum x_j^2 y_j)/(\sum x_j^4) = (195.0)/(61.1875) = 3.1869$ and the model $y = kx^2$ becomes $y = 3.1869x^2$.

Let's assume we prefer the model, $y = ax^2 + bx + c$.

$$xv := [.5, \quad 1, \quad 1.5, \quad 2, \quad 2.5];$$

$$yv := [.7, \quad 3.4, \quad 7.2, \quad 12.4, \quad 20.1].$$

The least squares fit, shown graphical in Figure 6.6, is

$$y = 3.2607x^2 - 0.22223x + 0.12630$$

Finding the parameters requires calculus or the use of excellent technology. We are only going to illustrate technology for this book and not the calculus development.

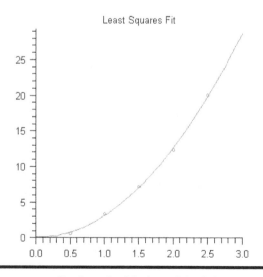

Figure 6.6   Least-squares fit plotted with the original data

6.4.2 Diagnostics and Interpretations

Coefficient of Determination: Statistical term: R^2.

In statistics, the **coefficient of determination**, R^2, is used in the context of statistical models whose main purpose is the prediction of future outcomes on the basis of other related information. It is the proportion of variability in a data set that is accounted for by the statistical model. It provides a measure of how well future outcomes are likely to be predicted by the model.

There are several different definitions of R^2, which are only sometimes equivalent. One class of such cases includes linear regression. In this case, R^2 is simply the square of the sample correlation coefficient between the outcomes and their predicted values, or in the case of simple linear regression, between the outcome and the values being used for prediction. In such cases, the values vary from 0 to 1. Important cases where the computational definition of R^2 can yield negative values, depending on the definition used, arise where the predictions which are being compared to the corresponding outcome have not derived from a model-fitting procedure using those data.

$R^2 = 1 - SSE/SST$

Values of R^2 outside the range 0 to 1 can occur where it is used to measure the agreement between observed and modeled values and where the "modeled" values are not obtained by linear regression and depending on which formulation of R^2 is used. If the first formula stated earlier is used, values can never be greater than one.

Residual Plots

In the previous section, you learned how to obtain a least squares fit of a model, and you plotted the model's predictions on the same graph as the observed data points in order to get a visual indication of how well the model matches the trend of the data. A powerful technique for quickly determining where the model breaks down or is adequate is to plot the actual deviations or residuals between the observed and predicted values as a function of the independent variable or the model. We plot the residuals in the y-axis and the model or the independent variable on the x-axis. The deviations should be randomly distributed and contained in a reasonably small band that is commensurate with the accuracy required by the model. Any excessively large residual warrants further investigation of the data point in question to discover the cause of the large deviation. A pattern or trend in the residuals indicates that a predictable effect remains to be modeled, and the nature of the pattern gives clues on how to refine the model, if a refinement is called for. We illustrate the possible patterns for residuals in Figures 6.7 (a)–(e). Our intent is to provide the modeler with knowledge concerning the adequacy of the model they have found. We will leave further investigations into correcting the patterns to follow on courses in statistical regression.

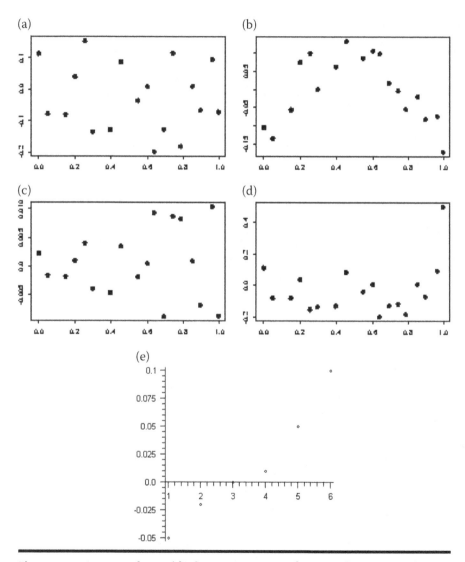

Figure 6.7 Patterns for residuals **(a)** no pattern **(b)** curved patter **(c)** fanning pattern **(d)** outliers **(e)** linear trend

6.4.3 Percent Relative Error

When using a model to predict information, we really want to know how well the model appears to work. We will use percent relative error (% REL ERR).

$$\% \operatorname{Re} l\ ERR = 100\% \cdot \frac{|y_{actual} - y_{model}|}{y_{actual}}$$

We really want these percent relative errors to be small (less than 10–20% on average).

Examples with Diagnostics and Inferential Statistics

Example 6.1: Wuhan Analysis Using Regression Techniques

We use our data originally shown in Table 1.6, Chapter 1, repeated now as Table 6.2.

We resent the data plot and an overlay of our model in Figure 6.8. Over the domain of the data used, the model looks excellent.

We start with fitting an exponential model, $y = ae^{bx}$

Our equation is

$712.268728868938 * e^{(.168334887822958\ *\ t)}$

We see a good fit in the given equation. We check a few diagnostics.

R^2 is quite good, as $R^2 = 0.98598$.

The residual plot, Figure 6.9, appears to have a trend, especially to the right where it dips and then rises.

Again looking at the data, we can find a death rate of approximately 4% (Figure 6.10.)

We might try to sue the model anyway. If we assume no changes in rates of infection or deaths, we can predict out 100 days. Then, we calculate the following:

Infections ($t = 100$) = 14,560,000,000 and the number of deaths, which is a function of the number infected, is 58,263,310,490. Again, this number is a lot larger than the entire population.

There are possible issues here with exponential growth models. Just because the computer gave us an answer, it does not mean the answer is correct. Here, our estimate exceeds the total population of Wuhan.

The model, although it has some good diagnostics, fails to pass the common sense test.

This suggests a refinement to the model. This is a great example of looking closely at our results and suggesting we need some changes to our model.

Therefore, perhaps we should have used a logistics model. A logistics model is a curve with a script *S* shape. The growth starts slowly and then increases more rapidly and slows down as it approaches the maximum population.

We apply a logistics model,

Table 6.2   Infections and Deaths, Wuhan, China

Time	Infections	Deaths
1	425	17
2	495	23
3	572	38
4	618	45
5	698	63
6	1,590	85
7	1,905	104
8	2,261	129
9	2,639	159
10	3,215	192
11	4,109	240
12	5,142	270
13	6,383	300
14	8,351	371
15	10,117	432
16	11,618	535
17	13,603	602
18	14,982	665
19	16,902	738
20	19,558	810

$$0.432073879135261e - 3/(1.46081684846028 * 10^{\wedge}(-8)$$
$$+ 1.58069176709821 * 10^{\wedge}(-6) * \exp(-.264715364083052 * t)).$$

The R^2 is still above 98% but the residual plot still has trends.

If we go to predict out 100 days, we get a more reasonable answer, infections = 29,578. We can assume this value is due to restrictions placed on the country and region. The number of deaths would be estimated as 4* of 29,578 or 1,184 people.

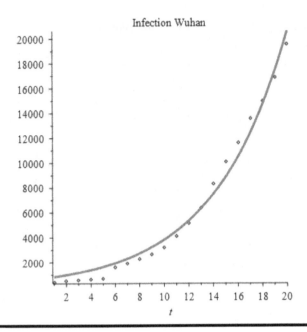

Figure 6.8   Data and model

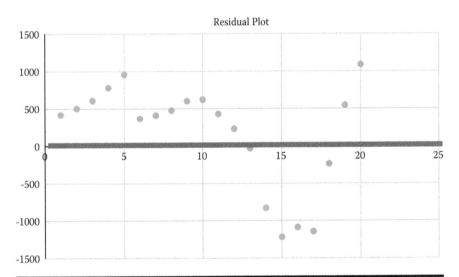

Figure 6.9   The residual plot from our exponential model

We plot the function and the data, see Figure 6.11.

We overlay the exponential model, the data, and the logistics model, see Figure 6.12.

We have added prediction intervals to the logistics equation developed. The plot is shown in Figure 6.13.

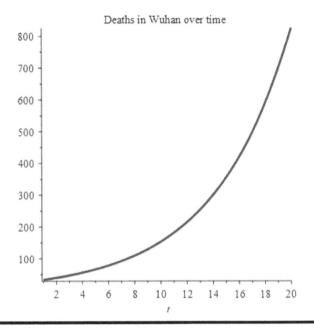

Figure 6.10   Death in Wuhan based on 4% of infections

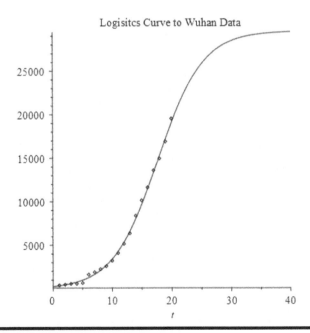

Figure 6.11   Logistics model to our Wuhan data

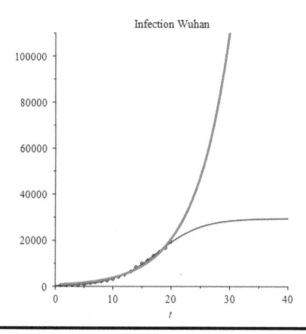

Figure 6.12   Overall of data and the two models: exponential as the upper curve, and logistics as the lower S shaped curve.

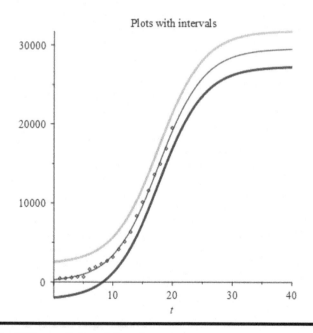

Figure 6.13   With prediction intervals

The results for prediction are anywhere between the upper and lower curves. The equation itself might need remodeling to include changes in the infection rate as it changes almost daily.

6.5 Chapter Summary

We have seen probabilistic models and regression models used to attempt to explain and predict the COVID-19 virus and its resulting data collected. Indicators here suggest that in both probabilistic models and regression models, "social distancing" work.

Chapter 7

Hypothesis Tests

Researchers use hypothesis testing to attempt to prove a stated claim at some level of statistical significance. In inferential statistics, there is a set procedure to do this testing.

Generally, there is a research question – this question could be, "Does the malaria drug work in combating COVID-19?"

Set up the hypothesis.

Set up a research design plan.

Collect the data.

Analyze the data.

Interpret the results.

Once you have identified the research question you need to answer, it is time to frame a good hypothesis. The hypothesis is the starting point for biostatistics and is usually based on a theory. Experiments are then designed to test the hypothesis. What is a hypothesis?

A research hypothesis is a statement describing a relationship between two or more variables that can be tested. A good hypothesis will be clear, avoid moral judgments, specific, objective, and relevant to the issue at hand. The design plans must be random trials of patients or test subjects. If there are more than one group for comparison, subjects must be randomly selected into each group. Then, we need a design plan. This is where we make all tests independent and identical. Usually any discrepancy here would void the subjects' results from being used in the analysis. After the data are collected, the data are used in the analysis, in this case, a hypothesis test, which will be described soon. We use this test to determine if we fail to reject the null hypothesis or reject the null hypothesis. Our goal in the test is to reject the null hypothesis at some key level of statistical significance. Generally speaking, there are type I and type II errors to be considered. Type I is the chance that a null hypothesis that is true is rejected and a type II error is when a null hypothesis is false and is not rejected. A final key aspect in biostatistical inference is the P-value.

7.1 The P-Value Approach to Test Hypothesis

Once the hypothesis has been designed, statistical tests help you to decide if you should accept or reject the null hypothesis. Statistical tests determine the p-value associated with the research data. The p-value is the probability that one could have obtained the result by chance; assuming the null hypothesis (H_0) was true.

We would write a P-value for a one tail test as $P\ (Z > |z_{statistic}|)$ or $P\ (T > |t_{statistic}|)$ and for a two tail test, $2 * P\ (Z > |z_{statistic}|)$ or $2 * P\ (T > |t_{statistic}|)$.

You must reject the null hypothesis if the P-value of the data falls below the predetermined level of statistical significance. Usually, the level of statistical significance is set at 0.05. If the P-value is less than 0.05, then you would reject the null hypothesis stating that there is no relationship between the predictor and the outcome in the population from which the sample is drawn.

However, if the P-value is greater than the predetermined level of significance, then there is *no statistically significant* association between the predictor and the outcome variable. This does not mean that there is no association between the predictor and the outcome in the population. It only means that the difference between the relationship observed and the relationship that could have occurred by random chance is small.

For example, null hypothesis (H_0): *The patients who take the study drug after a heart attack did not have a better chance of not having a second heart attack over the next 24 months.*

Data suggests that those who did not take the study drug were twice as likely to have a second heart attack with a P-value of 0.08. This p-value would indicate that there was an 8% chance that you would see a similar result (people on the placebo being twice as likely to have a second heart attack) in the general population because of random chance.

The hypothesis is not a trivial part of the clinical research process. It is a key element in a good biostatistics plan regardless of the clinical trial phase. There are many other concepts that are important for analyzing data from clinical trials.

Doing a hypothesis test is a detailed endeavor.

What is a hypothesis statement?

If you are going to propose a hypothesis, it's customary to write a statement. Your statement will look like this,

"If I ... (do this to an independent variable) .... then (this will happen to the dependent variable)."

For example:

- If I (give a patient a dose of the touted malaria drug) then (the patient has an improved status).
- If I (give patients counseling in addition to medication) then (their overall depression scale will decrease).
- If I (give COVID-19 testing to more people) then (we can get better control of the virus).
- If I (give the test subject a certain vaccine) then (they will become immune to the virus.

A good hypothesis statement should:

- Include an "if" and "then" statement (according to the University of California)
- Include both the independent and dependent variables
- Be testable by experiment, survey, or other scientifically sound technique
- Be based on information in prior research (either yours or someone else's)
- Have design criteria (for engineering or programming projects).

Now let's describe a hypothesis test.

A more powerful technique for inferring information about a parameter is a hypothesis test. A statistical hypothesis test is a claim about a single population characteristic or about values of several population characteristics. There is a null hypothesis (which is the claim initially favored or believe to be true) and is denoted by H_0. The other hypothesis, the alternate hypothesis, is denoted as H_a. We will always keep equality with the null hypothesis. The objective is to decide, based upon sample information, which of the two claims is correct. Typical hypothesis tests can be categorized by three cases:

Case 1:	H_0:	$\mu = \mu_0$	versus	H_a:	$\mu \neq \mu_0$
Case 2:	H_0:	$\mu \leq \mu_0$	versus	H_a:	$\mu > \mu_0$
Case 3:	H_0:	$\mu \geq \mu_0$	versus	H_a:	$\mu < \mu_0$

There are two types of errors that can be made in hypothesis testing, Type 1 errors called α error and Type II errors called β errors. It is important to understand these. Consider the information provided in the Table 7.1 below.

Some important facts about both α and β:

1. $\alpha = P$ (reject $H_0|H_0$ is true) $= P$ (Type I error)
2. $\beta = P$ (fail to reject $H_0|H_0$ is false) $= P$ (Type II error)
3. α is the level of significance of the test
4. $1 - \beta$ is the power of the test.

Thus, referring to the table, we would like α to be small, since it is the probability that we reject H_0 when H_0 is true. We would also want $1 - \beta$ to be large because it represents the probability that we reject H_0 when H_0 is false. Part of the modeling process is to determine which of these errors is the most costly and work to control that error as your primary error of interest.

An example is if you think of a trial where the person on trial is either guilty or innocent and the jury finds the person either guilty or not guilty We can think of hypothesis testing in the same context as a criminal trial in the United States. A criminal trial in the United States is a familiar situation in which a choice between two contradictory claims must be made.

1. The accuser of the crime must be judged either guilty or not guilty.
2. Under the US system of justice, the individual on trial is initially presumed not guilty.
3. Only STRONG EVIDENCE to the contrary causes the not guilty claim to be rejected in favor of a guilty verdict.
4. The phrase "beyond a reasonable doubt" is often used to set the cutoff value for when enough evidence has been given to convict.

Theoretically, we should never say, "The person is innocent," but instead, "There is not sufficient evidence to show that the person is guilty."

Now, let's compare that to how we look at a hypothesis test.

Table 7.1   Hypothesis Test

State of nature			
		H_0 *true*	H_a *true*
Test conclusion	Fail to reject H_0	$1 - \alpha$	β
	Reject H_0	α	$1 - \beta$

1. The decision about the population parameter(s) must be judged to follow one of two hypotheses.
2. We initially assume that H_0 is true.

The null hypothesis H_0 will be rejected (in favor of H_a) only if the sample evidence strongly suggests that H_0 is false. If the sample does not provide such evidence, H_0 will not be rejected.

The analogy to "beyond a reasonable doubt" in hypothesis testing is what is known as the **significance level**. This will be set before conducting the hypothesis test and is denoted as α.

We say we would rather let a guilty person go free than put an innocent person in jail.

Common values that we use for α are 0.1, 0.01, and 0.05.

State of nature			
		H_0 true	H_a true
Test conclusion	Fail to reject H_0	$1 - \alpha$	β
	Reject H_0	α	$1 - \beta$

The following is provided for hypothesis testing:

STEP 1: Identify the parameter of interest.
STEP 2: Determine the null hypothesis, H_0.
STEP 3: State the alternative hypothesis, H_a.
STEP 4: Give the formula for the test statistic based upon the assumptions that are satisfied.
STEP 5: State the rejection criteria based upon the value of α.
STEP 6: Obtain your sample data and substitute into your test statistic.
STEP 7: Determine the region in which your test statistics lies (rejection region or fail to reject region).
STEP 8: Make your statistical conclusion. Your choices are to either reject the null hypothesis or fail to reject the null hypothesis. Insure the conclusion is scenario oriented.
Step 9: Compute and interpret the P-Value of the test statistic.

Now, let's see some examples. In most hypothesis cases in biostatistics, our sample sizes will be quite large (much greater than 30) that allows us to do a Z-test.

7.2 Large Samples with σ is Unkown

Procedure:

1. Read the problem and recognize the claim.
2. Based upon the claim, called u_o, in (1) set up the test with H_o and H_a and μ. Choices are (a) two-tail test: Ho: $μ = u_o$ Ha: $μ ≠ u_o$ (b) right-tail test Ho: $μ ≤ u_o$ Ha: $μ > u_o$ (c) left-tail test Ho: $μ ≥ u_o$ Ha: $μ < u_o$. Notice that equality is in every H_o.
3. Draw the sketch to determine the rejection region (RR) and the fail to reject region (FTRR).

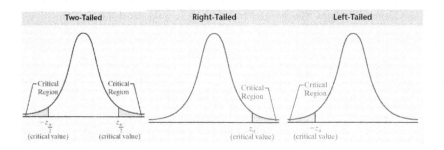

The values of $t_{a/2,\ df}$ and $t_{a,\ df}$ vary based upon a and df.
4. Calculate the Z value for the sample.
   For the sample, we need x-bar, s, and n. Use them to compute:

$$Z \ or \ t = \frac{(\bar{x} - \mu_0)}{\left(\frac{s}{\sqrt{n}}\right)}$$

where μ_0 is the equality claim value about the mean.
If you look at tables for the t distribution, as the degrees of freedom gets larger $t → Z$. Regardless, we will use the t distribution here.
5. Compare the sample t to the t values found in step 3 to see if our t statistic falls in the RR or the FTRR.
6. State conclusion.
7. Compute P-Value for the Sample t value as $2*P\ (T > |t|)$ for a two tail test and $P\ (T > |t|)$ for the other one tail tests. Reject the null hypothesis, Ho, if the P-Value is less than α. The P-value is the stronger argument.

Since no hypothesis tests concerning the COVID-19 has been published, we present some other medical, biostatistical, tests to illustrate how the tests work.

Example 7.1: Blood Glucose

Blood glucose levels for obese patients have a mean of 105 with a standard deviation of 15. A researcher thinks that a diet high in raw cornstarch will have a positive or negative effect on blood glucose levels. A sample of 36 patients who have tried the new diet have a mean glucose level of 140. Test the hypothesis that the new diet had an effect on glucose levels.

Step 1: State the null hypothesis: H_0: $\mu = 105$.

Step 2: State the alternate hypothesis: H_1: $\neq 105$.

Step 3: State your alpha level. We'll use 0.05 for this example. As this is a two-tailed test, split the alpha into two.

$0.05/2 = 0.025$

Step 4: Find the z-score associated with your alpha level. You're looking for the area in *one tail only*. A z-score for $0.75(1 - 0.025 = 0.975)$ is 1.96. As this is a two-tailed test, you would also be considering the left tail ($z = 1.96$).

Step 5: Find the test statistic using this formula: $Z = \frac{\bar{x} - \mu_0}{\sigma / \sqrt{n}}$

$$z = (140 - 105)/(15/\sqrt{36}) = 14.00$$

Step 6: If Step 5 is less than -1.96 or greater than 1.96 (Step 3), reject the null hypothesis. In this case, it is greater, so you *can* reject the null.

Step 7. Find the P-Value, $P(Z > |14.)) = 8.129 \times 10^{-45} \approx 0$.

We conclude that the new diet does affect glucose levels.

Example 7.2: Cholesterol Levels

Consider again the NCHS-reported mean total cholesterol level in 2002 for all adults of 203. Suppose a new drug is proposed to lower total cholesterol. A study is designed to evaluate the efficacy of the drug in lowering cholesterol. Fifteen patients are enrolled in the study and asked to take the new drug for six weeks. At the end of six weeks, each patient's total cholesterol level is measured and the sample statistics are as follows: $n = 100$, $\bar{x} = 198.9$, and $s = 28.5$. Is there statistical evidence of a reduction in mean total cholesterol in patients after using the new drug for six weeks? We will run the test using the five-step approach.

Step 1. Set up hypotheses and determine level of significance.

$$H_0: \ \mu = 203 \quad H_1: \ \mu < 203 \quad \alpha = 0.05$$

Step 2. Select the appropriate test statistic.

Because the sample size is large ($n > 30$), the appropriate test statistic is

$$Z \ or \ t = \frac{(\bar{x} - \mu_0)}{\left(\frac{s}{\sqrt{n}}\right)}$$

Step 3. Set up decision rule.

This is a lower tailed test, using a t statistic and a 5% level of significance. In order to determine the critical value of t, we need degrees of freedom, df, defined as $df = n - 1$. In this example $df = 100 - 1 = 99$. The critical value for a lower tailed test with $df = 99$ and $a = 0.05$ is -1.984 and if you used Z it would be -1.959 and the decision rule is as follows: Reject H_0 if $t \leq -2.145$ or $Z < -1.959$.

Step 4. Compute the test statistic.

We now substitute the sample data into the formula for the test statistic identified in Step 2.

Step 5. Conclusion.

We do not reject H_0 because $-14.38 < -1.984$ or -1.959. We do have statistically significant evidence at $\alpha = 0.05$ to show that the mean total cholesterol level is lower than the national mean in patients taking the new drug for six weeks. Again, because we rejected the null hypothesis, we make a stronger claim concluding statement allowing for the possibility that we may have committed a Type II error (i.e., failed to reject H_0 when in fact the drug is efficacious).

7.2.1 Large Samples with Proportions

To use this test, $p \ (1-p) \ n \geq 10$
   Procedure:

1. Read the problem and recognize the claim.
2. Based upon the claim, called p_o, in (1) set up the test with H_o and H_a and μ. Choices are (a) two-tail test: Ho: $p = p_o$ Ha: $p \neq p_o$ (b) right-tail test Ho: $p \leq p_o$ Ha: $p > p_o$ (c) left-tail test Ho: $p \geq p_o$ Ha: $p < p_o$. Notice that equality is in every H_o.
3. Draw the sketch to determine the rejection region (RR) and the fail to reject region (FTRR).

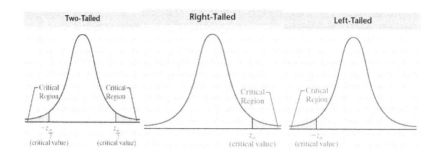

For α = 0.05

$$Z_{a/2} = \pm 1.96 \quad Z_a = 1.645 \quad Z_a = -1.645$$

For α = 0.01

$$Z_{a/2} = \pm 2.58 \quad Z_a = 2.33 \quad Z_a = -2.33$$

4. Calculate the Z value for the sample and with it compute if the test can be done using p-hat*(1-phat)*n must be greater than or equal to 10. If true (greater than or equal to 10) then we continue.
   For the sample we need p-hat (the sample proportion), $s = \sqrt{\frac{p_0(1-p_0)}{n}}$, and n. Use them to compute:

$$Z = \frac{(phat - \mu_0)}{\left(\sqrt{\frac{p_0(1-p_0)}{n}} \right)}$$

where p_0 is the equality claim value about the mean and phat is the sample proportion from the data,

5. Compare the sample Z to the Z values found in step 3 to see if our Z falls in the rejection region or the FTRR.
6. State conclusion.
7. Compute P-Value for the Sample Z value as 2*P (z > |Z|) for a two tail test and P (z > |Z|) for the other one tail tests. Reject the null hypothesis, Ho, if the P-Value is less than α. The P-value is the stronger argument for the conclusion and interpretation.

Example 7.3: Infection Rate

One argument going on today deals with the comparison of the COVID-19 pandemic and the Spanish flu pandemic in 1918. In our data search we found the world population was 1.8 billion (approximately) and that 50 million were infected (approximately). This makes the percent infected as 0.0277 or 2.7%.

How does this compare to today's pandemic in 2020. Our hypothesis says that this pandemic is not as bad as the Spanish flu pandemic in 1918.

Step 1. Set up the hypothesis test and choose level of significance as 0.05.

Ho: P = 0.0277

Ha: P > 0.0277 (worst case is that this pandemic is worse)

Step 2. Check to make sure we can use the test.

If we use US data our population is 334 million and we have 1.5 million infected as of May 6, 2020, phat = 0.004 (1–P) = 0.996, n = 334,000,000 so we find the product is about 1.5 million.

$$z = \frac{\hat{p} - p_0}{\sqrt{\frac{p_0(1 - p_0)}{n}}}$$

$$z = \frac{0.044 - 0.0277}{\sqrt{\frac{0.0277(1 - 0.0277)}{334000000}}} = -.139$$

Since –.139 is not > –1.959 we fail to reject the null hypothesis.

We remind the reader that the Spanish flu is over the percent and numbers were after two years of the virus. The US data are after about five months of the virus in this country.

Example 7.4: Death Rates

One argument going on today deals with the comparison of the COVID-19 pandemic and the Spanish flu pandemic in 1918 in terms of deaths. In our data search we found the world population at 1.8 billion (approximately) and that 50 million were infected (approximately). This makes the percentage infected as 0.0277 or 2.7%. Of the 50 million infected, 17.4 million died. This is a death rate of 34.8%.

How does this compare to today's death rate for the pandemic in 2020. Our hypothesis says that the death rate for this pandemic is not as bad as the Spanish flu pandemic in 1918.

Step 1. Set up the hypothesis test and choose level of significance as 0.05.

Ho: P = 0.348

Ha: P > 0.348 (worst case is that this pandemic is worse)

Step 2. Check to make sure we can use the test.

If we use US data our population is 334 million and we have 1.5 million infected as of May 6, 2020 and 80,000 deaths, which is 5.33%, phat = 0.0533 (1−0.0533) = 0.9467, n = 1.5 million so we find the product is about 75,733 > 10.

$$z = \frac{\hat{p} - p_0}{\sqrt{\frac{p_0(1 - p_0)}{n}}}$$

$$z = \frac{0.0533 - 0.348}{\sqrt{\frac{(.348)(1 - .348)}{3340000001500000}}} = -757$$

Since −757 is not >1.959 we fail to reject the null hypothesis.

Chapter 8

Two Samples Hypothesis Test (Means and Proportions)

In the field of medical testing, the design plans may have two study groups: one getting the actual drug and the other getting a placebo. The claim is usually that in the study group getting the drug has better outcomes. There might be more survivors, faster recovery time or rate, and so on.

Currently testing is ongoing for vaccines and treatments. This data are proprietary and will not be used. We will use more generic examples to illustrate the procedures that might be used. It is important to realize and keep in mind that everything we said about testing a single sample still holds.

Let's set the stage with rules for hypothesis testing of two samples:

Although we are interested in if $\mu_1 = \mu_2$, we will test the difference, $\mu_1 - \mu_2 = 0$.
If we consider that $\mu_1 - \mu_2 =$ "constant," we can test that as well.

8.1 Large Independent Samples (Both Sample Sizes $m, n \geq 30$)

To test hypothesis about the difference of the means from two large independent samples, we will follow the procedure as follows.

Tests comparing two sample means from sample X and sample Y
H_o: $\mu_1 = \mu_2$ we write this as $\mu_1 - \mu_2 = 0$
H_a: This can be any of the following as required:

$$\Delta\mu \neq 0, \quad \Delta\mu < 0 \quad \Delta\mu > 0$$

Test Statistic: $Z = \dfrac{(\bar{x} - \bar{y})}{\sqrt{\frac{s_1^2}{m} + \frac{s_2^2}{n}}}$

Decision: Reject the claim, Ho if and only if (iff) for
$\Delta\mu \neq 0$ Either $Z \geq z_{\alpha/2}$ or $Z \leq -z_{\alpha/2}$

$$\Delta\mu < 0 \quad Z \leq -z_\alpha$$

$$\Delta\mu > 0 \quad Z \geq z_\alpha$$

Calculate the P-value
   Two-tailed test, $P = 2 * P\ (Z > |Z|)$
   One-tail test, $p = P\ (Z > |z|)$

8.2 Large Independent Samples [Only One Sample Size (Either *m* or *n*) Is ≥30]

The procedure here is different as now we use a t-distribution.
   H_o: $\mu_1 = \mu_2$ we write this as $\mu_1 - \mu_2 = 0$
   H_a: This can be any of the following as required:
$\Delta\mu \neq 0, \quad \Delta\mu < 0 \quad \Delta\mu > 0$

Test Statistic: $t = \dfrac{(\bar{x} - \bar{y}) - (\mu_1 - \mu_2)}{S_p \sqrt{\frac{1}{m} + \frac{1}{n}}}$

Where $S_p^2 = \dfrac{(m-1)s_1^2 + (n-1)s_2^2}{m+n-2}$

And $S_p = \sqrt{S_p^2}$

Decision: Reject the claim, Ho iff for
$\Delta\mu \neq 0$ Either $t \geq t_{\alpha/2}$ or $t \leq -t_{\alpha/2}$

$$\Delta\mu < 0 \quad t \leq -t_\alpha$$

$$\Delta\mu > 0 \quad t \geq t_\alpha$$

and for df use $m + n - 2$.

Paired Data
   Often when the lengths of the data are the same m = n, then we might use paired data.

A sampling method is **independent** when the individuals selected for one sample do not dictate which individuals are to be in a second sample. A sampling method is **dependent** when the individuals selected to be in one sample are used to determine the individuals to be in the second sample. Dependent samples are often referred to as **matched pairs** samples.

Step 1. A claim is made regarding the mean of the difference, d. We let X_1 represent the first sample and X_2 represent the 2nd sample, then d $= X_1 - X_2$. The claim is used to determine the null hypothesis and alternative hypothesis as before.

Two-tail	Left-tail	Right tail
Ho: $\mu_d = 0$	Ho: $\mu_d \geq 0$	Ho: $\mu_d \leq 0$
Ha: $\mu_d \neq 0$	Ha: $\mu_d < 0$	Ha: $\mu_d > 0$

Step 2. Select a level of significance a based upon the serious of making a type I error. The level of significance is usually 0.05 or 0.01. To obtain the critical value we use $|t_{a/2, \text{ df}}|$ for two tail test or $|t_{a, \text{ df}}|$ for a one tail test.

Step 3. Compute the test statsitic, t.

$$t = \frac{\bar{d} - 0}{\frac{s_d}{\sqrt{n}}}$$

Which approximately follows a student t distribution with n–1 degree of freedom. The values of $\bar{d}$ and s_d are the mean and standard deviation of the difference, $d = X_1 - X_2$

Step 4. Compare the test statistics to the t critical value.

Step 5. Make a decision.

Step 6. Compute the P-value.

Two tail test 2*P (T > |t|)

One tail test P (T > |t|)

8.2.1 Two Sample Tests on Proportions

These days in 2020, we might be concerned with the testing for COVID-19 vaccines. Most tests have a placebo for one sample and the "real" drug for the other samples. We will want lots of people in each sample for statistical relevance. The claim will be that number of protected people in the "real" drug sample is higher than the placebo sample. So, how will we test this? Table 8.1 provides a framework.

To test if our test will be valid, we must see if $p_1(1-p_1)\,n$ and $p_2(1-p_2)\,m$ are each greater than or equal to 10, that insures normality of our proportions.

Test about $p_1 - p_2$ for two independent samples.

	Two-tail test	Right-tail test	Left-tail test
Step 1.	Ho: $p_1 - p_2 = 0$	Ho: $p_1 - p_2 = 0$	Ho: $p_1 - p_2 = 0$
	Hα = : $p_1 - p_2 \neq 0$	>Hα = : $p_1 - p_2 > 0$	Hα = : $p_1 - p_2 < 0$
Step 2.	Choose α =	Choose α =	Choose α =
Step 3.	Test statistic	Test statistic	Test statistic
	$z = \dfrac{p_1 - p_2}{\sqrt{\hat{p}\hat{q}\left(\frac{1}{m}+\frac{1}{n}\right)}}$	$z = \dfrac{p_1 - p_2}{\sqrt{\hat{p}\hat{q}\left(\frac{1}{m}+\frac{1}{n}\right)}}$	$z = \dfrac{p_1 - p_2}{\sqrt{\hat{p}\hat{q}\left(\frac{1}{m}+\frac{1}{n}\right)}}$
Step 4.	*Determine rejection Regions*	Determine rejection regions.	*Determine rejection regions.*
	$|z| > Za_{za/2}$	$z \geq z_a$	$z \leq z_a$
Step 5.	Compare test statistic to rejection region, make decision.	Compare test statistic to rejection region, make decision.	Compare test statistic to rejection region, make decision.
Step 6.	Compute P-Value and make decision	Compute P-Value and make decision	Compute P-Value and make decision
	P = 2*minimum (P(Z $\geq$ |z|))	P = (P(Z $\geq$ z))	P = P(Z $\leq$ z)

8.2.2 Tests with Two Independent Samples, Continuous Outcome

There are many applications where it is of interest to compare two independent groups with respect to their mean scores on a continuous outcome. Here, we compare means between groups, but rather than generating an estimate of the difference, we will test whether the observed difference (increase, decrease or difference) is statistically significant or not. Remember, that hypothesis testing gives an assessment of statistical significance, whereas estimation gives an estimate of effect and both are important.

Here, we discuss the comparison of means when the two comparison groups are independent or physically separate. The two groups might be determined by a particular attribute (e.g., sex, diagnosis of cardiovascular disease) or might be set up by the investigator (e.g., participants assigned to receive an experimental treatment or placebo). The first step in the analysis involves computing descriptive statistics on each of the two samples. Specifically, we compute the sample size, mean and standard deviation in each sample and we denote these summary statistics as follows:

For sample 1:

$$n1$$
$$\bar{X}_1$$
$$s1$$

For sample 2:

$$n2$$
$$\bar{X}_2$$
$$s2$$

The designation of sample 1 and sample 2 is arbitrary. In a clinical trial setting the convention is to call the treatment group 1 and the control group 2. However, when comparing men and women, for example, either group can be 1 or 2.

In the two independent samples application with a continuous outcome, the parameter of interest in the test of hypothesis is the difference in population means, $\mu_1-\mu_2$. The null hypothesis is always that there is no difference between groups with respect to means, i.e.,

$$H_0: \mu_1 - \mu_2 = 0$$

The null hypothesis can also be written as follows: $H_0: \mu_1 = \mu_2$. In the research hypothesis, an investigator can hypothesize that the first mean is larger than the second ($H_1: \mu_1 > \mu_2$), that the first mean is smaller than the second ($H_1: \mu_1 < \mu_2$), or that the means are different ($H_1: \mu_1 \neq \mu_2$). The three different alternatives represent upper-, lower-, and two-tailed tests, respectively. The following test statistics are used to test these hypotheses.

Test Statistics for Testing $H_0: \mu_1 = \mu_2$

- if $n_1 \geq 30$ and $n_2 \geq 30$

$$z = \frac{\bar{X}_1 - \bar{X}_2}{S_p\sqrt{\frac{1}{n_1} + \frac{1}{n_2}}}$$

- if $n_1 < 30$ or $n_2 < 30$

$$t = \frac{\bar{X}_1 - \bar{X}_2}{S_p\sqrt{\frac{1}{n_1} + \frac{1}{n_2}}}$$

where $df = n_1 + n_2 - 2$.

NOTE: The formulas above assume equal variability in the two populations (i.e., the population variances are equal, or $s_1^2 = s_2^2$). This means that the outcome is equally variable in each of the comparison populations. For analysis, we have samples from each of the comparison populations. If the sample variances are similar, then the assumption about variability in the populations is probably reasonable. As a guideline, if the ratio of the sample variances, s_1^2/s_2^2 is between 0.5 and 2 (i.e., if one variance is no more than double the other), then the formulas above are appropriate. If the ratio of the sample variances is greater than 2 or less than 0.5 then alternative formulas must be used to account for the heterogeneity in variances.

The test statistics include Sp, which is the pooled estimate of the common standard deviation (again assuming that the variances in the populations are similar) computed as the weighted average of the standard deviations in the samples as follows:

$$S_p = \sqrt{\frac{(n_1 - 1)s_1^2 + (n_2 - 1)s_2^2}{n_1 + n_2 - 2}}$$

Because we are assuming equal variances between groups, we pool the information on variability (sample variances) to generate an estimate of the variability in the population. Note: Because Sp is a weighted average of the standard deviations in the sample, Sp will always be in between s_1 and s_2.)

Example 8.1: Blood Pressure between Men and Women

Data measured on $n = 3,539$ participants who attended the seventh examination of the Children in the Framingham Heart Study are shown below. The data of interest is as follows

	Men	Women
Sample size, n	1,623	1,911
Systolic blood pressure mean	128.2	126.5
Systolic blood pressure standard deviation	17.5	20.1

Suppose we now wish to assess whether there is a statistically significant difference in mean systolic blood pressures between men and women using a 5% level of significance.

Step 1. Set up hypotheses and determine level of significance

$$H_0: \mu_1 = \mu_2$$

$$H_1: \mu_1 \neq \mu_2 \quad \alpha = 0.05$$

Step 2. Select the appropriate test statistic.

Because both samples are large (≥ 30), we can use the Z test statistic as opposed to t. Note that statistical computing packages use t throughout. Before implementing the formula, we first check whether the assumption of equality of population variances is reasonable. The guideline suggests investigating the ratio of the sample variances, s_1^2/s_2^2. Suppose we call the men group 1 and the women group 2. Again, this is arbitrary; it only needs to be noted when interpreting the results. The ratio of the sample variances is $17.5^2/20.1^2 = 0.76$, which falls between 0.5 and 2 suggesting that the assumption of equality of population variances is reasonable. The appropriate test statistic is

$$z = \frac{\bar{X}_1 - \bar{X}_2}{S_p\sqrt{\frac{1}{n_1} + \frac{1}{n_2}}}.$$

Step 3. Set up decision rule.

This is a two-tailed test, using a Z statistic and a 5% level of significance. Reject H_0 if Z ≤ -1.960 or is Z ≥ 1.960.

Step 4. Compute the test statistic.

We now substitute the sample data into the formula for the test statistic identified in Step 2. Before substituting, we will first compute Sp, the pooled estimate of the common standard deviation.

$$S_p = \sqrt{\frac{(n_1 - 1)s_1^2 + (n_2 - 1)s_2^2}{n_1 + n_2 - 2}}$$

$$S_p = \sqrt{\frac{(1623 - 1)17.5^2 + (1911 - 10)20.1^2}{1623 + 1911 - 2}} = \sqrt{359.12} = 19.0$$

Notice that the pooled estimate of the common standard deviation, Sp, falls in between the standard deviations in the comparison groups (i.e., 17.5 and 20.1). Sp is slightly closer in value to the standard deviation in the women (20.1) as there were slightly more women in the sample. Recall, Sp is a weight average of the standard deviations in the comparison groups, weighted by the respective sample sizes.

Now the test statistic:

$$Z = \frac{128.2 - 126.5}{19.0\sqrt{\frac{1}{162.3} + \frac{1}{1911}}} = \frac{1.7}{0.64} = \mathbf{2.66}$$

Step 5. Conclusion.

We reject H_0 because $2.66 \geq 1.960$. We have statistically significant evidence at $\alpha = 0.05$ to show that there is a difference in mean systolic blood pressures between men and women. The p-value is $p < 0.010$.

Here again we find that there is a statistically significant difference in mean systolic blood pressures between men and women at $p < 0.010$. Notice that there is a very small difference in the sample means ($128.2 - 126.5 = 1.7$ units), but this difference is beyond what would be expected by chance. Is this a clinically meaningful difference? The large sample size in this example is driving the statistical significance. A 95% confidence interval for the difference in mean systolic blood pressures is: 1.7 ± 1.26 or $(0.44, 2.96)$. The confidence interval provides an assessment of the magnitude of the difference between means whereas the test of hypothesis and p-value provide an assessment of the statistical significance of the difference.

Above we performed a study to evaluate a new drug designed to lower total cholesterol. The study involved one sample of patients, each patient took the new drug for six weeks and had their cholesterol measured. As a means of evaluating the efficacy of the new drug, the mean total cholesterol following six weeks of treatment was compared to the NCHS-reported mean total cholesterol level in 2002 for all adults of 203. At the end of the example, we discussed the appropriateness of the fixed comparator as well as an alternative study design to evaluate the effect of the new drug involving two treatment groups, where one group receives the new drug and the other does not.

Example 8.2: Lowering Total Cholesterol

A new drug is proposed to lower total cholesterol. A randomized controlled trial is designed to evaluate the efficacy of the medication in lowering cholesterol. Thirty participants are enrolled in the trial and are randomly assigned to receive either the new drug or a placebo. The participants do not know which treatment they are assigned. Each participant is asked to take the assigned treatment for six weeks. At the end of six weeks, each patient's total cholesterol level is measured and the sample statistics are as follows.

Treatment	Sample Size	Mean	Standard Deviation
New Drug	15	195.9	28.7
Placebo	15	227.4	30.3

Is there statistical evidence of a reduction in mean total cholesterol in patients taking the new drug for six weeks as compared to participants taking placebo? We will run the test using the five-step approach.

Step 1. Set up hypotheses and determine level of significance

$$H_0: \mu_1 = \mu_2 \quad H_1: \mu_1 < \mu_2 \quad \alpha = 0.05$$

Step 2. Select the appropriate test statistic.

Because both samples are small (< 30), we use the t test statistic. Before implementing the formula, we first check whether the assumption of equality of population variances is reasonable. The ratio of the sample variances, $s_1^2/s_2^2 = 28.7^2/30.3^2 = 0.90$, which falls between 0.5 and 2, suggesting that the assumption of equality of population variances is reasonable. The appropriate test statistic is:

$$t = \frac{\bar{X}_1 - \bar{X}_2}{S_p \sqrt{\frac{1}{n_1} + \frac{1}{n_2}}}.$$

Step 3. Set up decision rule.

This is a lower-tailed test, using a t statistic and a 5% level of significance. The appropriate critical value can be found in the t Table (in More Resources to the right). In order to determine the critical value of t we need degrees of freedom, df, defined as $df = n_1 + n_2 - 2 = 15 + 15 - 2 = 28$. The critical value for a lower tailed test with $df = 28$ and $\alpha = 0.05$ is −1.701 and the decision rule is: Reject H_0 if $t \leq -1.701$.

Step 4. Compute the test statistic.

We now substitute the sample data into the formula for the test statistic identified in Step 2. Before substituting, we will first compute Sp, the pooled estimate of the common standard deviation.

$$S_p = \sqrt{\frac{(n_1 - 1)s_1^2 + (n_2 - 1)s_2^2}{n_1 + n_2 - 2}}$$

$$S_p = \sqrt{\frac{(15 - 1)28.7^2 + (15 - 1)30.3^2}{15 + 15 - 2}} = \sqrt{870.89} = 29.5$$

Now the test statistic,

$$t = \frac{195.9 - 227.4}{29.5\sqrt{\frac{1}{15} + \frac{1}{15}}} = \frac{-31.5}{10.77} = -2.92$$

Step 5. Conclusion.

We reject H_0 because $-2.92 \leq -1.701$. We have statistically significant evidence at $\alpha = 0.05$ to show that the mean total cholesterol level is lower in patients taking the new drug for six weeks as compared to patients taking placebo, $p < 0.005$.

The clinical trial in this example finds a statistically significant reduction in total cholesterol, whereas in the previous example where we had a historical control (as opposed to a parallel control group) we did not demonstrate efficacy of the new drug. Notice that the mean total cholesterol level in patients taking placebo is 217.4 which is very different from the mean cholesterol reported among all Americans in 2002 of 203 and used as the comparator in the prior example. The historical control value may not have been the most appropriate comparator as cholesterol levels have been increasing over time. In the next section, we present another design that can be used to assess the efficacy of the new drug.

8.2.3 Tests with Matched Samples, Continuous Outcome

In the previous section we compared two groups with respect to their mean scores on a continuous outcome. An alternative study design is to compare matched or paired samples. The two comparison groups are said to be **dependent,** and the data can arise from a single sample of participants where each participant is measured twice (possibly before and after an intervention) or from two samples that are matched on specific characteristics (e.g., siblings). When the samples are dependent, we focus on **difference scores** in each participant or between members of a pair and the test of hypothesis is based on the mean difference, μ_d. The null hypothesis again reflects "no difference" and is stated as H_0: $\mu_d = 0$. Note that there are some instances where it is of interest to test whether there is a difference of a particular magnitude (e.g., $\mu_d = 5$) but in most instances the null hypothesis reflects no difference (i.e., $\mu_d = 0$).

The appropriate formula for the test of hypothesis depends on the sample size. The formulas are shown below and are identical to those we presented for estimating the mean of a single sample presented (e.g., when comparing against an external or historical control), except here we focus on difference scores.

Test Statistics for Testing H_0: $\mu_d = 0$

If $n \geq 30$

$$z = \frac{\bar{X}_d - \mu_d}{s_d / \sqrt{n}}$$

If n < 30

$$t = \frac{\bar{X}_d - \mu_d}{s_d / \sqrt{n}}$$

where df = n − 1

Example 8.3: Drug Test

A new drug is proposed to lower total cholesterol and a study is designed to evaluate the efficacy of the drug in lowering cholesterol. Fifteen patients agree to participate in the study and each is asked to take the new drug for six weeks. However, before starting the treatment, each patient's total cholesterol level is measured. The initial measurement is a pre-treatment or baseline value. After taking the drug for six weeks, each patient's total cholesterol level is measured again and the data are shown below. The rightmost column contains difference scores for each patient, computed by subtracting the six-week cholesterol level from the baseline level. The differences represent the reduction in total cholesterol over 4 weeks. (The differences could have been computed by subtracting the baseline total cholesterol level from the level measured at six weeks. The way in which the differences are computed does not affect the outcome of the analysis, and only the interpretation.)

Subject Identification Number	*Baseline*	*Six Weeks*	*Difference*
1	215	205	10
2	190	156	34
3	230	190	40
4	220	180	40
5	214	201	13
6	240	227	13
7	210	197	13
8	193	173	20
9	210	204	6
10	230	217	13

(Continued)

Subject Identification Number	Baseline	Six Weeks	Difference
11	180	142	38
12	260	262	−2
13	210	207	3
14	190	184	6
15	200	193	7

Because the differences are computed by subtracting the cholesterols measured at six weeks from the baseline values, positive differences indicate reductions and negative differences indicate increases (e.g., participant 12 increases by 2 units over 6 weeks). The goal here is to test whether there is a statistically significant reduction in cholesterol. Because of the way in which we computed the differences, we want to look for an increase in the mean difference (i.e., a positive reduction). In order to conduct the test, we need to summarize the differences. In this sample, we have the following statistics:

$$n = 15$$
$$\bar{x} = 16.9333$$
$$s_d = 14.2$$

Is there statistical evidence of a reduction in mean total cholesterol in patients after using the new medication for six weeks? We will run the test using the five-step approach.

■ **Step 1.** Set up hypotheses and determine level of significance

$$H_0: \mu_d = 0 \quad H_1: \mu_d > 0 \quad \alpha = 0.05$$

NOTE: If we had computed differences by subtracting the baseline level from the level measured at six weeks then negative differences would have reflected reductions and the research hypothesis would have been $H_1: \mu_d < 0$.

■ **Step 2.** Select the appropriate test statistic.

Because the sample size is small ($n < 30$) the appropriate test statistic is

$$t = \frac{\bar{X}_d - \mu_d}{s_d / \sqrt{n}}.$$

■ **Step 3.** Set up decision rule.

This is an upper-tailed test, using a t statistic and a 5% level of significance. The appropriate critical value can be found in the t Table at the right, with df = 15 − 1 = 14. The critical value for an upper-tailed test with df = 14 and α = 0.05 is 2.145 and the decision rule is Reject H_0 if t ≥ 2.145.

■ **Step 4.** Compute the test statistic.
We now substitute the sample data into the formula for the test statistic identified in Step 2.

$$t = \frac{\bar{X}_d - \mu_d}{s_d/\sqrt{n}} = \frac{16.9 - 0}{14.2/\sqrt{15}} = 4.61$$

■ **Step 5.** Conclusion.
We reject H_0 because 4.61 ≥ 2.145. We have statistically significant evidence at α = 0.05 to show that there is a reduction in cholesterol levels over six weeks.

Here we illustrate the use of a matched design to test the efficacy of a new drug to lower total cholesterol. We also considered a parallel design (randomized clinical trial) and a study using a historical comparator. It is extremely important to design studies that are best suited to detect a meaningful difference when one exists. There are often several alternatives and investigators work with biostatisticians to determine the best design for each application. It is worth noting that the matched design used here can be problematic in that observed differences may only reflect a "placebo" effect. All participants took the assigned medication, but is the observed reduction attributable to the medication or a result of these participation in a study.

8.2.4 Tests with Two Independent Samples, Proportions

Here we consider the situation where there are two independent comparison groups and the outcome of interest is dichotomous (e.g., success/failure). The goal of the analysis is to compare proportions of successes between the two groups. The relevant sample data are the sample sizes in each comparison group (n_1 and n_2) and the sample proportions ($\hat{p}_1$ and $\hat{p}_2$) which are computed by taking the ratios of the numbers of successes to the sample sizes in each group, i.e.,

$$\hat{p}_1 = \frac{x_1}{n_1} \text{ and } \hat{p}_2 = \frac{x_2}{n_2}$$

There are several approaches that can be used to test hypotheses concerning two independent proportions. Here we present one approach - the chi-square test of independence is an alternative, equivalent, and perhaps more popular approach to the same analysis.

In tests of hypothesis comparing proportions between two independent groups, one test is performed and results can be interpreted to apply to a risk difference, relative risk or odds ratio. As a reminder, the risk difference is computed by taking the difference in proportions between comparison groups, the risk ratio is computed by taking the ratio of proportions, and the odds ratio is computed by taking the ratio of the odds of success in the comparison groups. Because the null values for the risk difference, the risk ratio and the odds ratio are different, the hypotheses in tests of hypothesis look slightly different depending on which measure is used. When performing tests of hypothesis for the risk difference, relative risk or odds ratio, the convention is to label the exposed or treated group 1 and the unexposed or control group 2.

For example, suppose a study is designed to assess whether there is a significant difference in proportions in two independent comparison groups. The test of interest is as follows:

$$H_0: p_1 = p_2 \text{ versus } H_1: p_1 \neq p_2.$$

The following are the hypothesis for testing for a difference in proportions using the risk difference, the risk ratio and the odds ratio. First, the hypotheses above are equivalent to the following:

- For the risk difference, $H_0: p_1 - p_2 = 0$ versus $H_1: p_1 - p_2 \neq 0$ which are, by definition, equal to $H_0: RD = 0$ versus $H_1: RD \neq 0$.
- If an investigator wants to focus on the risk ratio, the equivalent hypotheses are $H_0: RR = 1$ versus $H_1: RR \neq 1$.
- If the investigator wants to focus on the odds ratio, the equivalent hypotheses are $H_0: OR = 1$ versus $H_1: OR \neq 1$.

Suppose a test is performed to test $H_0: RD = 0$ versus $H_1: RD \neq 0$ and the test rejects H_0 at $\alpha = 0.05$. Based on this test we can conclude that there is significant evidence, $\alpha = 0.05$, of a difference in proportions, significant evidence that the risk difference is not zero, significant evidence that the risk ratio and odds ratio are not one. The risk difference is analogous to the difference in means when the outcome is continuous. Here the parameter of interest is the difference in proportions in the population, $RD = p_1 - p_2$ and the null value for the risk difference is zero. In a test of hypothesis for the risk difference, the null hypothesis is always $H_0: RD = 0$. This is equivalent to $H_0: RR = 1$ and $H_0: OR = 1$. In the research hypothesis, an investigator can hypothesize that the first proportion is larger than the second ($H_1: p_1 > p_2$, which is equivalent to $H_1: RD > 0$, $H_1: RR > 1$ and $H_1: OR > 1$), that the first proportion is smaller than the second ($H_1: p_1 < p_2$, which is equivalent to $H_1: RD < 0$, $H_1: RR < 1$ and $H_1: OR < 1$), or that the proportions are different ($H_1: p_1 \neq p_2$, which is equivalent to $H_1: RD \neq 0$, $H_1: RR \neq 1$ and $H_1: OR \neq 1$). The three different alternatives represent upper-, lower- and two-tailed tests, respectively.

The formula for the test of hypothesis for the difference in proportions is given below.

Test Statistics for Testing H0: $p_1 = p$

$$z = \frac{\hat{p}_1 - \hat{p}_2}{\sqrt{\hat{p}(1 - \hat{p})\left(\dfrac{1}{n_1} + \dfrac{1}{n_2}\right)}}$$

Where $\hat{p}_1$ is the proportion of successes in sample 1, $\hat{p}_2$ is the proportion of successes in sample 2, and $\hat{p}$ is the proportion of successes in the pooled sample. $\hat{p}$ is computed by summing all of the successes and dividing by the total sample size, as follows: $p = \frac{x_1 + x_2}{n_1 + n_2}$ (this is similar to the pooled estimate of the standard deviation, Sp, used in two independent samples tests with a continuous outcome; just as Sp is in between s_1 and s_2, $\hat{p}$ will be in between $\hat{p}_1$ and $\hat{p}_2$).

The formula above is appropriate for large samples, defined as at least 5 successes ($np \geq 5$) and at least 5 failures ($n(1-p \geq 5)$) in each of the two samples. If there are fewer than 5 successes or failures in either comparison group, then alternative procedures, called exact methods must be used to estimate the difference in population proportions.

Example 8.4: The following table summarizes data from n = 3,799 participants who attended the fifth examination of the Children in the Framingham Heart Study. The outcome of interest is prevalent CVD and we want to test whether the prevalence of CVD is significantly higher in smokers as compared to non-smokers.

	Free of CVD	History of CVD	Total
Non-Smoker	2,757	298	3,055
Current Smoker	663	81	744
Total	3,420	379	3,799

The prevalence of CVD (or proportion of participants with prevalent CVD) among non-smokers is 298/3,055 = 0.0975 and the prevalence of CVD among current smokers is 81/744 = 0.1089. Here smoking status defines the comparison groups and we will call the current smokers group 1 (exposed) and the nonsmokers (unexposed) group 2. The test of hypothesis is conducted below using the five step approach.

■ **Step 1.** Set up hypotheses and determine level of significance

$$H_0: p_1 = p_2 \quad H_1: p_1 \neq p_2 \quad \alpha = 0.05$$

■ **Step 2.** Select the appropriate test statistic.

We must first check that the sample size is adequate. Specifically, we need to ensure that we have at least 5 successes and 5 failures in each comparison group. In this example, we have more than enough successes (cases of prevalent CVD) and failures (persons free of CVD) in each comparison group. The sample size is more than adequate so the following formula can be used:

$$z = \frac{\hat{p}_1 - \hat{p}_2}{\sqrt{\hat{p}(1 - \hat{p})\left(\frac{1}{n_1} + \frac{1}{n_2}\right)}}.$$

■ **Step 3.** Set up decision rule.

Reject H_0 if $Z \le -1.960$ or if $Z \ge 1.960$.

■ **Step 4.** Compute the test statistic.

We now substitute the sample data into the formula for the test statistic identified in Step 2. We first compute the overall proportion of successes:

$\hat{p} = \frac{x_1 + x_2}{n_1 + n_2} = \frac{81 + 298}{744 + 3055} = \frac{379}{3799} = 0.0998$

We now substitute to compute the test statistic.

$$z = \frac{\hat{p}_1 - \hat{p}_2}{\sqrt{\hat{p}(1 - \hat{p})\left(\frac{1}{n} + \frac{1}{n}\right)}} = \frac{0.1089 - 0.0975}{\sqrt{0.0988(1 - 0.988)\left(\frac{1}{744} + \frac{1}{3055}\right)}} = \frac{0.0114}{0.0123}$$

$= 0.927$

■ **Step 5.** Conclusion.

We do not reject H_0 because $-1.960 < 0.927 < 1.960$. We do not have statistically significant evidence at $\alpha = 0.05$ to show that there is a difference in prevalent CVD between smokers and non-smokers.

A 95% confidence interval for the difference in prevalent CVD (or risk difference) between smokers and non-smokers as 0.0114 ± 0.0247, or between -0.0133 and 0.0361. Because the 95% confidence interval for the risk difference includes zero we again conclude that there is no statistically significant difference in prevalent CVD between smokers and non-smokers.

Smoking has been shown over and over to be a risk factor for cardiovascular disease. What might explain the fact that we did not observe a statistically significant difference using data from the Framingham Heart Study? HINT: Here we consider prevalent CVD, would the results have been different if we considered incident CVD?

Example 8.5: Effectiveness of a Pain Reliever

A randomized trial is designed to evaluate the effectiveness of a newly developed pain reliever designed to reduce pain in patients following joint replacement surgery. The trial compares the new pain reliever to the pain reliever currently in use (called the standard of care). A total of 100 patients undergoing joint replacement surgery agreed to participate in the trial. Patients were randomly assigned to receive either the new pain reliever or the standard pain reliever following surgery and were blind to the treatment assignment. Before receiving the assigned treatment, patients were asked to rate their pain on a scale of 0–10 with higher scores indicative of more pain. Each patient was then given the assigned treatment and after 30 minutes was again asked to rate their pain on the same scale. The primary outcome was a reduction in pain of 3 or more scale points (defined by clinicians as a clinically meaningful reduction). The following data were observed in the trial.

Treatment Group	n	Number with Reduction of 3+ Points	Proportion with Reduction of 3+ Points
New Pain Reliever	50	23	0.46
Standard Pain Reliever	50	11	0.22

We now test whether there is a statistically significant difference in the proportions of patients reporting a meaningful reduction (i.e., a reduction of 3 or more scale points) using the five step approach.

■ **Step 1.** Set up hypotheses and determine level of significance

$$H_0: p_1 = p_2 \quad H_1: p_1 \neq p_2 \quad \alpha = 0.05$$

Here the new or experimental pain reliever is group 1 and the standard pain reliever is group 2.

■ **Step 2.** Select the appropriate test statistic.

We must first check that the sample size is adequate. Specifically, we need to ensure that we have at least 5 successes and 5 failures in each comparison group, i.e.,

$$\min(n_1 \hat{p}_1, \quad n_1(1 - \hat{p}_1), \quad n_2 \hat{p}_2, \quad n_2(1 - \hat{p}_2)) \geq 5$$

In this example, we have min (50(0.46), 50(1–0.46), 50(0.22), 50(1–0.22)) = min (23, 27, 11, 39) = 11. The sample size is adequate so the following formula can be used

$$z = \frac{\hat{p}_1 - \hat{p}_2}{\sqrt{\hat{p}(1 - \hat{p})\left(\frac{1}{n_1} + \frac{1}{n_2}\right)}}$$

- **Step 3.** Set up decision rule.
  Reject H_0 if $Z \leq -1.960$ or if $Z \geq 1.960$.
- **Step 4.** Compute the test statistic.
  We now substitute the sample data into the formula for the test statistic identified in Step 2. We first compute the overall proportion of successes:

$$\hat{p} = \frac{x_1 + x_2}{n_1 + n_2} = \frac{23 + 11}{50 + 50} = \frac{34}{100} = 0.34$$

We now substitute to compute the test statistic.

$$z = \frac{\hat{p}_1 - \hat{p}_2}{\sqrt{\hat{p}(1 - \hat{p})\left(\frac{1}{n_1} + \frac{1}{n_2}\right)}} = \frac{0.46 - 0.22}{\sqrt{0.34(1 - 0.34)\left(\frac{1}{50} + \frac{1}{50}\right)}} = \frac{0.24}{0.095} = 2.526$$

- **Step 5.** Conclusion.
  We reject H_0 because $2.526 \geq 1.960$. We have statistically significant evidence at a = 0.05 to show that there is a difference in the proportions of patients on the new pain reliever reporting a meaningful reduction (i.e., a reduction of 3 or more scale points) as compared to patients on the standard pain reliever.

A 95% confidence interval for the difference in proportions of patients on the new pain reliever reporting a meaningful reduction (i.e., a reduction of 3 or more scale points) as compared to patients on the standard pain reliever is 0.24 ± 0.18 or between 0.06 and 0.42. Because the 95% confidence interval does not include zero we concluded that there was a statistically significant difference in proportions which is consistent with the test of hypothesis result.

8.3 Chapter Summary

Here we presented hypothesis testing techniques for means and proportions in one- and two-sample situations. Tests of hypothesis involve several steps, including specifying the null and alternative or research hypothesis, selecting and computing an appropriate test statistic, setting up a decision rule and drawing a conclusion. There are many details to consider in hypothesis testing. The first is to determine the appropriate test. We discussed Z and t tests here for different applications. The appropriate test depends on the distribution of the outcome variable (continuous or dichotomous), the number of comparison groups (one, two), and whether the comparison groups are independent or dependent. The following table summarizes the different tests of hypothesis discussed here.

- Continuous Outcome, One Sample: H0: $\mu = \mu 0$

$$z = \frac{\bar{X} = \mu_0}{s/\sqrt{n}}$$

- Continuous Outcome, Two Independent Samples: H0: $\mu 1 = \mu 2$

$$z = \frac{\bar{X}_1 - \bar{X}_2}{s_p \sqrt{\frac{1}{n_1} + \frac{1}{n_2}}} \text{ and } s_p = \sqrt{\frac{(n_1 - 1)s_1^2 + (n_2 - 1)s_2^2}{n_1 + n_2 - 2}}$$

- Continuous Outcome, Two Matched Samples: H0: $\mu d = 0$

$$z = \frac{\bar{X}_d - \mu_d}{s_d/\sqrt{n}}$$

and $s_d = \sqrt{\dfrac{\Sigma 22\, \text{Differences}^2 - (\Sigma\, \text{Differences})^2 / n}{n - 1}}$

- Dichotomous Outcome, One Sample: H0: $p = p0$

$$z = \frac{\hat{p} - p_0}{\sqrt{\frac{p_0(1 - p_0)}{n}}}$$

- Dichotomous Outcome, Two Independent Samples: H0: $p1 = p2$, RD = 0, RR = 1, OR = 1

$$z = \frac{\hat{p}_1 - \hat{p}_2}{\sqrt{\hat{p}(1 - \hat{p})\left(\frac{1}{n_1} + \frac{1}{n_2}\right)}}$$

Once the type of test is determined, the details of the test must be specified. Specifically, the null and alternative hypotheses must be clearly stated. The null hypothesis always reflects the "no change" or "no difference" situation. The alternative or research hypothesis reflects the investigator's belief. The investigator might hypothesize that a parameter (e.g., a mean, proportion, difference in means or proportions) will increase, will decrease, or will be different under specific conditions (sometimes the conditions are different experimental conditions and other times the conditions are simply different groups of participants). Once the hypotheses are specified, data are collected and summarized. The appropriate test is then conducted according to the five-step approach. If the test leads to rejection of the null hypothesis, an approximate p-value is computed to summarize the significance of the findings. When tests of hypothesis are conducted using statistical computing packages, exact p-values are computed. Because the statistical tables in this textbook are limited, we can only approximate p-values. If the test fails to reject the null hypothesis, then a weaker concluding statement is made for the following reason.

In hypothesis testing, there are two types of errors that can be committed. A Type I error occurs when a test incorrectly rejects the null hypothesis. This is referred to as a false positive result, and the probability that this occurs is equal to the level of significance, α. The investigator chooses the level of significance in Step 1, and purposely chooses a small value such as $\alpha = 0.05$ to control the probability of committing a Type I error. A Type II error occurs when a test fails to reject the null hypothesis when in fact it is false. The probability that this occurs is equal to β. Unfortunately, the investigator cannot specify β at the outset because it depends on several factors including the sample size (smaller samples have higher b), the level of significance (β decreases as α increases), and the difference in the parameter under the null and alternative hypothesis.

We noted in several examples in this chapter, the relationship between confidence intervals and tests of hypothesis. The approaches are different, yet related. It is possible to draw a conclusion about statistical significance by examining a confidence interval. For example, if a 95% confidence interval does not contain the null value (e.g., zero when analyzing a mean difference or risk difference, one when analyzing relative risks or odds ratios), then one can conclude that a two-sided test of hypothesis would reject the null at $\alpha = 0.05$. It is important to note that the correspondence between a confidence interval and test of hypothesis relates to a two-sided test and that the confidence level corresponds to a specific level of significance (e.g., 95% to $\alpha = 0.05$, 90% to $\alpha = 0.10$ and so on). The exact significance of the test, the p-value, can only be determined using the hypothesis testing approach and the p-value provides an assessment of the strength of the evidence and not an estimate of the effect.

As vaccine tests continue to develop, they will eventually go into trials. In these trials, then, potential vaccines will be compared to some other drug or placebo to see if in fact they measure success with a statistical significance. This will be the only way of measuring success before any vaccine is mass produced and disseminated to the public.

Chapter 9

Agent-Based Model with NetLogo

NetLogo has a self-contained agent-based model. One is called the *epidi model* that we will run under various scenarios.

After looking at the setup, it appears to be more suitable to smaller communities rather than cities, states, or countries. However, it is helpful to examine it as a useful model of measuring the effects of an agent (person infected).

Each scenario is shown in the figures with the parameters provided to the model. To begin with, let's state what NetLogo says about this model:

> This model simulates the spread of an infectious disease in a closed population. It is an introductory model in the curricular unit called epiDEM (Epidemiology: Understanding Disease Dynamics and Emergence through Modeling). This particular model is formulated based on a mathematical model that describes the systemic dynamics of a phenomenon that emerges when one infected person is introduced in a wholly susceptible population. This basic model, in mathematical epidemiology, is known as the Kermack-McKendrick model.

The Kermack-McKendrick model assumes a closed population, meaning there are no births, deaths, or travel into or out of the population. It also assumes that there is homogeneous mixing, in that each person in the world has the same chance of interacting with any other person within the world. In terms of the virus, the model assumes that there are no latent or dormant periods, nor a chance of viral mutation.

Because this model is so simplistic in nature, it facilitates mathematical analyses and also the calculation of the threshold at which an epidemic is expected to occur. We call this the reproduction number, and denote it as R_0. Simply, R_0 stands for the number of secondary infections that arise as a result of introducing one infected person in a wholly susceptible population, over the course of the infected person's contagious period (i.e., while the person is infective, which, in this model, is from the beginning of infection until recovery).

This model incorporates all of the aforementioned assumptions, but each individual has a 5% chance of being initialized as infected. This model shows the disease spread as a phenomenon with an element of stochasticity. Small perturbations in the parameters included here can in fact lead to different final outcomes.

Overall, this model helps users to:

1. Engage in a new way of viewing/modeling epidemics that is more personable and relatable
2. Understand how the reproduction number, R_0, represents the threshold for an epidemic
3. Think about different ways to calculate R_0, and the strengths and weaknesses in each approach
4. Understand the relationship between derivatives and integrals, represented simply as rates and cumulative number of cases, and
5. Provide opportunities to extend or change the model to include some properties of a disease that interest users the most.

How It Works

Individuals wander around the world in random motion. Upon coming into contact with an infected person, by being in any of the eight surrounding neighbors of the infected person or in the same location, an uninfected individual has a chance of contracting the illness. The user sets the number of people in the world, as well as the probability of contracting the disease.

An infected person has a probability of recovering after reaching their recovery time period, which is also set by the user. The recovery time of each individual is determined by pulling from an approximately normal distribution with a mean of the average recovery time set by the user.

The colors of the individuals indicate the state of their health. Three colors are used: white individuals are uninfected, red individuals are infected, and green individuals are recovered. Once recovered, the individual is permanently immune to the virus.

The graph infection and recovery rates shows the rate of change of the cumulative infected and recovered in the population. It tracks the average number of secondary infections and recoveries per tick. The reproduction number is calculated under different assumptions than those of the Kermack-McKendrick model, as we allow for more than one infected individual in the population, and introduce aforementioned variables.

At the end of the simulation, the R_0 reflects the estimate of the reproduction number, the final size relation that indicates whether there will be (or there was, in the model sense) an epidemic. This again closely follows the mathematical derivation that R_0 = beta*S (0)/ gamma* = *N*ln(S (0)/S (t))/(N – S (t)), where N is the total population, S (0) is the initial number of susceptibles, and S (t) is the total number of susceptibles at time t. In this model, the R_0 estimate is the number of secondary infections that arise for an average infected individual over the course of the person's infected period.

How To Use It

The SETUP button creates individuals according to the parameter values chosen by the user. Each individual has a 5% chance of being initialized as infected. Once the model has been setup, push the GO button to run the model. GO starts the model and runs it continuously until GO is pushed again.

Note that in this model each time-step can be considered to be in hours, although any suitable time unit will do.

What follows is a summary of the sliders in the model.

INITIAL-PEOPLE (initialized to vary between 50 and 400): The total number of individuals in the simulation, determined by the user.

INFECTION-CHANCE (10–100): Probability of disease transmission from one individual to another.

RECOVERY-CHANCE (10–100): Probability of an infected individual to recover once the infection has lasted longer than the person's recovery time.

AVERAGE-RECOVERY-TIME (50–300): The time it takes for an individual to recover on average. The actual individual's recovery time is pulled from a normal distribution centered around the AVERAGE-RECOVERY-TIME at its mean, with a standard deviation of a quarter of the AVERAGE-RECOVERY-TIME. Each time-step can be considered to be in hours, although any suitable time unit will do.

A number of graphs are also plotted in this model.

CUMULATIVE INFECTED AND RECOVERED: This plots the total percentage of infected and recovered individuals over the course of the disease spread.

POPULATIONS: This plots the total number of people with or without the flu over time.

INFECTION AND RECOVERY RATES: This plots the estimated rates at which the disease is spreading. BetaN is the rate at which the cumulative infected changes, and Gamma rate at which the cumulative recovered changes.

R_0: This is an estimate of the reproduction number, only comparable to the Kermack McKendrick's definition if the initial number of infected were 1.

With this explanation, we begin our modeling and sensitivity analysis.

See for the model itself:

Yang and Wilensky (2011). NetLogo epiDEM Basic model. http://ccl. northwestern.edu/netlogo/models/epiDEMBasic. Center for Connected Learning and Computer-Based Modeling, Northwestern University, Evanston, IL.

9.1 Scenario 1

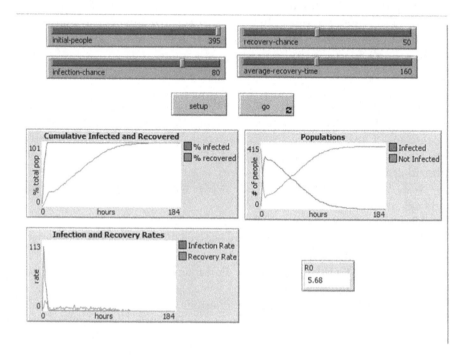

We call this our standard scenario. The parameters chosen were initial people in the simulation, infection chance, recovery chance, and average recovery time. We see with these parameters that the infection dies away but we see the rates fluctuate as well.

We noted here that R0 was higher than a previous model we discussed, which was 2.75. We experimented with the slider until we came close with a Ro of 2.95. We decided to use this as our standard setting.

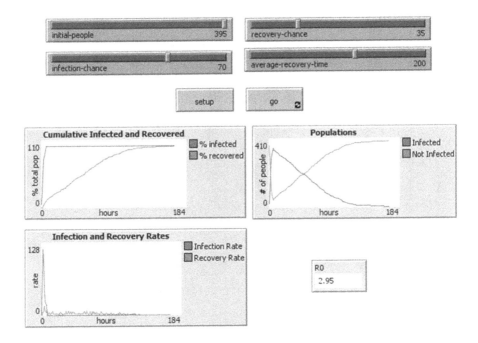

In this model, we have a closed population of 395 and infection rate as described as 70% in close quarters, recovery chance of 35%, and an average recovery time of 200. We see in this model that everyone gets infected.

We then do some sensitivity analysis by changing one or more parameters and seeing the effect on the results.

9.2 Scenario 2 Sensitivity Analysis One

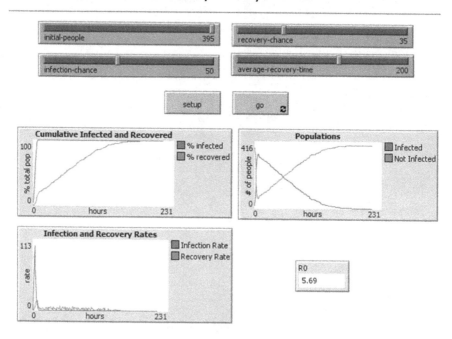

Let's change an input slightly; the infection rate decreases to 50%. Again, 100% get infected but it takes longer.

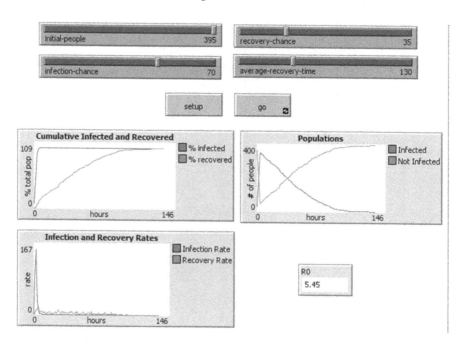

A decrease in the recovery time leads to everyone infected sooner.

Let's make several changes – recovery time decreases to 130 and recovery change to 30.

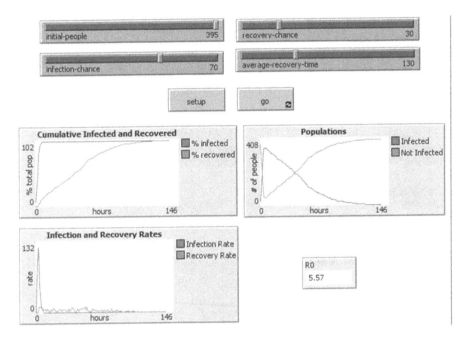

The model still shows a quick 100% infection rate.

The applicability of these models to explain COVID-19 appears to be small unless we were modeling the effects in a small school.

We watch carefully how the curves change. The infection rate and susceptible are similar to the curves we obtained in regression and SIR models.

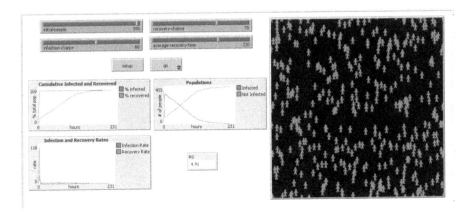

In each case, in these agent-based models, the infection dies out. We think this is a modeling artifact of the range and the parameters that can be changed to alter the model. When is it really bad? We don't know.

There is also a version of the model that considers travel to the hospital. We ran one trial case to see how it appears to work.

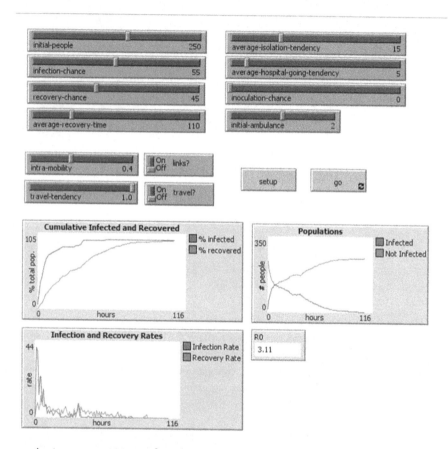

Again, we see 100% infected in about 100 plus hours. That seems a bit fast to achieve that rate.

Since the COVID-19 data change daily, we really do not know which model is best for predicting. We might only know which model is best for yesterday in only small contained locales.

Chapter 10

Concluding Remarks and Epilogue

In this book, we presented various model forms and results. The modeling process has shown us that the assumptions for the models are essential in the analysis. We hope we have presented interesting cases with real data and have shown the possible results based upon how well the model fits the data. We believe we have shown that social distancing works. It might not be six feet but it is working and most likely more than six feet would be better. It was discussed on the CBS nightly news on May 13, 2020 that one person infected 15 others at a club in Korea. Imagine that each of these 15 people infect only 2 more, and then those infect 2 more. It does not take long to be back at an exponential growth model. That is not what we need to beat this pandemic. We suggest patience and following the social guidelines until everyone can be administered the vaccine. The alternative of trying to reach herd immunity would most likely lead to millions of deaths.

If the models are wrong and the virus disappears all by itself, we will know it and the only cost has been time – the time to save lives.

Wolfe Blizter on CNN said in 2020 the model back then predicted 147,000 dead by August 2020. The prediction was based on loosening social distancing restrictions and people going out and mingling.

We provide a snapshot of the model's results as a reminder.

l to 6/6	Pure Projection 6/28 to 7/4			Pure Projection 7/26 to 8/1			Pure Proj
Sat Jun 6	0.56%	Sun Jun 28	Sat Jul 4	0.30%	Sun Jul 26	Sat Aug 1	0.15%
	0.56%			0.33%			0.15%
	5.32%			5.39%			5.20%
NA				NA			NA
NA				NA			NA
NA				NA			NA
330,908,505				330,908,505			330,908,505
7,283	0	wk dths	4,983	0		3,105	0
125,489	0.00643	wk cases	86,088	0.00261		53,721	0.00617

extrapolation	Predicted deaths with line fit extrapolation			Predicted deaths with line fit extrapolation			Predicted deaths
predicted new deaths per day	predicted total deaths	predicted daily death ratio	predicted new deaths per day	predicted total deaths	predicted daily death ratio	predicted new deaths per day	predicted total deaths
1,040	132,611	0.549%	712	147,656	0.304%	444	158,328
989	132,611	0.53%	675	147,656	0.26%	417	158,326
954	131,936	0.49%	647	147,255	0.27%	401	158,111
1,026	131,289	0.53%	698	146,838	0.30%	434	157,905
1,082	130,591	0.57%	739	146,404	0.32%	460	157,682
909	129,852	0.48%	625	145,944	0.27%	390	157,444
1,142	129,226	0.61%	787	145,554	0.34%	458	157,243
1,172	128,419	0.64%	812	145,063	0.35%	509	156,988

| extrapolation | Predicted cases with line fit extrapolation | | | Predicted cases with line fit extrapolation | | | Predicted cases |

Screenshots from [AlidadeOnlineDiscussion] [ext] Re: [external] Is there light or darkness at the end of the COVID-19 tunnel?, May 12, 2020).

Note the circled estimate. It is now July 13, 2020 as I am finalizing this book. We are very close to the prediction because many have become lax in their use of masks and social distancing. This does not include all the protests, most done without masks or social distancing, over the death of George Floyd.

Be safe! We can beat this together.

References

Chapter 1

Allman, E. and J. Rhodes, 2004. *Mathematical modeling with biology: An introduction.* Cambridge Press.

Albright, B. 2010. *Mathematical modeling with Excel.* Jones and Bartlett Publishers.

Bender, E. 2000. *Mathematical modeling.* Dover Press.

Fox, W. P. 2012. *Mathematical modeling with Maple.* Cengage Publishing.

Fox, W. 2013. Mathematical Modeling and Analysis: An example using a Catapult. *Computer in Education Journal,* **4**(3): 69–77.

Fox, W. 2014. Chapter 17, Game Theory in Business and Industry. *Encyclopedia of Business Analytics and Optimization.* IGI Global and Sage Publications. **V**(1): 162–173.

Fox, W. 2014. Chapter 221, TOPSIS in Business Analytics. *Encyclopedia of Business Analytics and Optimization.* IGI Global and Sage Publications. **V**(5): 281–291.

Fox, W. P. 2016. Applications and Modeling using multi-attribute decision making to rank terrorist threats. *Journal of Socialomics,* **5**(2), pp. 1–12.

Fox, W. 2018. Mathematical Modeling for Business Analytics. Taylor and Francis Publishers, Boca Raton, FL.

Fox, W. & W Bauldry. 2019. Problem Solving with Maple, Volume I. Taylor and Francis Publishers, Boca Raton, FL.

Fox, W. & W Bauldry. 2020. Problem Solving with Maple, Volume II. Taylor and Francis Publishers, Boca Raton, FL.

Giordano, F. R., W. Fox, and S. Horton. 2014. *A first course in mathematical modeling* (5th ed.). Brooks-Cole Publishers.

Myer, W. 2004. *Concepts of mathematical modeling.* Dover Press.

Meerschaert, M. M. 1999. *Mathematical modeling.* 2nd edition. Academic Press.

Saaty, T. 1980. *The analytical hierarchy process.* McGraw Hill.

Winston, W. 1995. *Introduction to mathematical programming.* Duxbury Press. pp. 323–325.

AidadeOnlineDiscussion May 6, 2020 (email, on line discussions, May 6–7, 2020) https://www.weforum.org/agenda/2020/05/compare-coronavirus-reponse-excess-deaths-rates/ (accessed May 1, 2020).

Chapter 2

Fox, W. P. 2012. *Mathematical modeling with Maple.* Cengage Publishing.

Fox, W. 2018. Mathematical Modeling for Business Analytics. Taylor and Francis Publishers, Boca Raton, FL.

Fox, W. & W. Bauldry. 2019. Problem Solving with Maple, Volume I. Taylor and Francis Publishers, Boca Raton, FL.

Fox, W. & W. Bauldry. 2020. Problem Solving with Maple, Volume II. Taylor and Francis Publishers, Boca Raton, FL.

Giordano, F. R., W. Fox, and S. Horton. 2014. *A first course in mathematical modeling* (5th ed.). Brooks-Cole Publishers.

https://www.worldometers.info/coronavirus/country/.

Bhatnager, M. COVID-19 Models and Predictions, preprint April 2020, **Preprint** · April 2020 DOI: 10.13140/RG.2.2.29541.96488, www.researchgate.net/publication/340375647_COVID-19_Mathematical_Modeling_and_Predictions.

Rodgers, K, Wolfe, J. and L Bronner, May 13, 2020, FiveThirtyEight, Science, (https://fivethirtyeight.com/features/without-a-vaccine-herd-immunity-wont-save-us/) accessed May 13, 2020/.

Chapter 3

Allman, E. and J. Rhoades 2004. *Mathematical Modeling with Biology: An introduction.* Cambridge Press.

Fox, W. and W. Bauldry, 2020. Problem Solving with Maple, Vol 1. Taylor and Francis Publishers. Boca Raton, FL.

Fox, W. and W. Bauldry, 2020. Problem Solving with Maple, Vol 2. Taylor and Francis Publishers. Boca Raton, FL.

Giordano, F., W. Fox, & S. Horton. 2014. *A first course in mathematical modeling*, 5th Ed. Cengage Publishing, Boston: MA.

Sandefur, J. 2003. *Elementary mathematical modeling.* Thompson Publishing.

Bonder, S. 1981. Mathematical Modeling of Military Conflict Situations, *Proceedings of Symposia in Applied Mathematics, Volume 25, Operations Research, Mathematics and Models*, American Mathematical Society.

Braun, M. 1981. *Differential Equations and Their Applications*, 3rd ed. Springer-Verlag.

Coleman, C. S. 1983. "Combat Models", in *Differential Equation Models*, Martin Braun, Courtney Coleman, and Donald Drew, Editors, Vol 1 of *Models in Applied Mathematics*, William Lucas, Editor. Springer-Verlag.

Lanchester, F. W. 1956. Mathematics in Warfare The World of Mathematics, J. Newman ed. Simon and Shuster. Vol. 4.

Teague, D. 2005. Combat Models, Teaching Contemporary Mathematics Conference.

Meerschaert, M. 1999. *Mathematical Modeling*, 2nd Edition. Academic Press.

Fox, W. P. 2012. *Mathematical modeling with Maple.* Cengage Publishing.

Fox, W. 2018. Mathematical Modeling for Business Analytics. Taylor and Francis Publishers, Boca Raton, FL.

Fox, W. & W Bauldry. 2019. Problem Solving with Maple, Volume I. Taylor and Francis Publishers, Boca Raton, FL.

Fox, W. & W Bauldry. 2020. Problem Solving with Maple, Volume II. Taylor and Francis Publishers, Boca Raton, FL.

Giordano, F. R., W. Fox, and S. Horton. 2014. *A first course in mathematical modeling* (5th ed.). Brooks-Cole Publishers.

https://www.worldometers.info/coronavirus/country/.

https://qz.com/1816060/a-chart-of-the-1918-spanish-flu-shows-why-social-distancing-works/.

Li, Y., Wang, B., Peng, R., Zhou, C., Zhan, Y., Liu, Z., Jiang, X., and Zhao, B. Mathematical Modeling and Epidemic Prediction of COVID-19 and Its Significance to Epidemic Prevention and Control Measures, http://www.remedypublications.com/open-access/mathematical-modeling-and-epidemic-prediction-of-covid-19-and-its-significance-5755.pdf, accessed May 10, 2020.

Chapter 4

Fox, W. P. 2012. *Mathematical modeling with Maple.* Cengage Publishing.

Fox, W. 2018. Mathematical Modeling for Business Analytics. Taylor and Francis Publishers, Boca Raton, FL.

Fox, W. & W. Bauldry. 2019. Problem Solving with Maple, Volume I. Taylor and Francis Publishers, Boca Raton, FL.

Fox, W. & W. Bauldry. 2020. Problem Solving with Maple, Volume II. Taylor and Francis Publishers, Boca Raton, FL.

Giordano, F. R., W. Fox, and S. Horton. 2014. *A first course in mathematical modeling* (5th ed.). Brooks-Cole Publishers.

https://www.worldometers.info/coronavirus/country/.

Weckesser, Warren. Colgate University notes posted on line (http://math.colgate.edu/~wweckesser/math312Spring05/handouts/PeriodicDrugDose.pdf).

Chapter 5

Fox, W. P. 2012. *Mathematical modeling with Maple.* Cengage Publishing.

Fox, W. 2018. Mathematical Modeling for Business Analytics. Taylor and Francis Publishers, Boca Raton, FL.

Fox, W. & W. Bauldry. 2019. Problem Solving with Maple, Volume I. Taylor and Francis Publishers, Boca Raton, FL.

Fox, W. & W. Bauldry. 2020. Problem Solving with Maple, Volume II. Taylor and Francis Publishers, Boca Raton, FL.

Giordano, F. R., W. Fox, and S. Horton. 2014. *A first course in mathematical modeling* (5th ed.). Brooks-Cole Publishers.

https://www.worldometers.info/coronavirus/country/.

Weckesser, W. (Colgate University), accessed May 9, 2020. (http://math.colgate.edu/~wweckesser/math312Spring05/handouts/PeriodicDrugDose.pdf).

Smith, D. and L. Moore, SIR Models. https://www.maa.org/press/periodicals/loci/joma/the-sir-model-for-spread-of-disease-the-differential-equation-model, accessed May 2, 2020).

Taubenberger, J. K. & Morens, D. M. (2006). 1918 Influenza: the Mother of All Pandemics. *Emerging Infectious Diseases* 12(1), pp. 15–22. https://doi.org/10.3201/eid1201.050979.

Jeffery K. Taubenberger* and David M. Morens[†] Author information Copyright and License information Disclaimer. https://www.ncbi.nlm.nih.gov/pmc/articles/PMC3291398/.

Yoneyanna, T. & M. Krishnamoothy, Spanish Flu, (accessed May 2, 2020). https://arxiv.org/ftp/arxiv/papers/1006/1006.0019.pdf.

COVID 19 projections Accessed May 2020, (https://covid19.healthdata.org/united-states-of-america).

COVID-19 Forecasts https://www.cdc.gov/coronavirus/2019-ncov/covid-data/forecasting-us.html.

Future of the Pandemic, NY Times https://www.nytimes.com/2020/05/08/health/coronavirus-pandemic-curve-scenarios.html.

G. Edenharter, SIR Model with Births and Deaths, Edenharter Research.\ (09/06/2015). Available online: https://www.edenharter-research.de/.

Chapter 6

https://www.worldometers.info/coronavirus/country/.
Cano, B. & S Morales. 2020. COVID-19 Modelling: the Effects of Social Distancing. doi: https://doi.org/10.1101/2020.03.29.20046870, https://www.medrxiv.org/content/10.1101/2020.03.29.20046870v1 (accessed May 15, 2020).

Chapter 7

https://www.worldometers.info/coronavirus/country/.
https://courses.lumenlearning.com/boundless-statistics/chapter/hypothesis-testing-two-samples/.
http://sphweb.bumc.bu.edu/otlt/MPH-Modules/BS/BS704_HypothesisTest-Means-Proportions/BS704_HypothesisTest-Means-Proportions6.html (example used but modified).
Wapole R. Myers, R., Myers, S. and K. Ye, Probability and Statistics for Engineers and Scientists, 9th Edition, Pearson Publishers, Boston, MA.
Devore, J. Probability and Statistics for Engineering and the Scientists, 8 ed. Cengage Publishers, Boston, Ma. 2015.
Sullivan, M. Fundamentals of Statistics, ed 5e. Pearson Publishing. Boston, MA 2018.
https://www.worldometers.info/coronavirus/country/.

Chapter 8

https://www.worldometers.info/coronavirus/country/.
https://courses.lumenlearning.com/boundless-statistics/chapter/hypothesis-testing-two-samples/.
http://sphweb.bumc.bu.edu/otlt/MPH-Modules/BS/BS704_HypothesisTest-Means-Proportions/BS704_HypothesisTest-Means-Proportions6.html example used.
Wapole R. Myers, R., Myers, S. and Ye K. (2012). Probability and Statistics for Engineers and Scientists, 9th Edition. Pearson Publishers, Boston, MA.
Devore, J. 2015. Probability and Statistics for Engineering and the Sciencest, 8 ed. Cengage Publsihers, Boston, Ma.
Sullivan, M. 2018. Fundamentals of Statisitcs, ed 5e. Pearson Publishing. Boston, MA.

Chapter 9

NetLogo https://ccl.northwestern.edu/netlogo/models/epiDEMBasic (accessed May 9, 2020).
Yang, C. and Wilensky, U. (2011). NetLogo epiDEM Basic model. http://ccl.northwestern.edu/netlogo/models/epiDEMBasic. Center for Connected Learning and Computer-Based Modeling, Northwestern University, Evanston, IL.
Wilensky, U. (1999). NetLogo. http://ccl.northwestern.edu/netlogo/. Center for Connected Learning and Computer-Based Modeling, Northwestern University, Evanston, IL.

Chapter 10

Model's Predictions, [AlidadeOnlineDiscussion] [ext] Re: [external] Is there light or darkness at the end of the COVID-19 tunnel?, email. https://www.fanniemae.com/research-and-insights/forecast/theres-light-end-covid-19-tunnel, May 12, 2020.

Index

Note: *Italicized* page numbers refer to figures, **bold** page numbers refer to tables